RANKING TASK
EXERCISES IN PHYSICS

edited by

Thomas L. O'Kuma
Lee College
Baytown, Texas

David P. Maloney
Indiana University-Purdue University Fort Wayne
Fort Wayne, Indiana

Curtis J. Hieggelke
Joliet Junior College
Joliet, Illinois

Prentice Hall, Upper Saddle River, New Jersey 07458

Library of Congress Cataloging-in-Publication Data

Ranking task exercises in physics / edited by Thomas L. O'Kuma, David P. Maloney, Curtis J. Hieggelke.
    p. cm .- - (Prentice Hall series in educational innovation)
   Includes bibliographical references
   ISBN 0-13-022355-7
   1. Physics Problems, exercises, etc. I. O'Kuma, Thomas L   II.
Maloney, David P.   III. Hieggelke, Curtis J.   IV. Series.
   QC32 .R28 1999
    530' 076 dc21   99-32915
                                         CIP

Executive Editor: Alison Reeves
Editor-in-Chief: Paul F. Corey
Assistant Vice President of Production and Manufacturing: David W. Riccardi
Executive Managing Editor: Kathleen Schiaparelli
Assistant Managing Editor: Lisa Kinne
Production Editor: Linda DeLorenzo
Marketing Manager: Steven Sartori
Editorial Assistant: Gillian Buonanno
Manufacturing Manager: Trudy Pisciotti
Art Director: Jayne Conte
Cover Designer: Bruce Kenselaar

10 9 8 7 6 5 4 3

ISBN 0-13-022355-7

Prentice-Hall International (UK) Limited, London
Prentice-Hall of Australia Pty. Limited, Sydney
Prentice-Hall Canada, Inc., Toronto
Prentice-Hall Hispanoamericana, S.A., Mexico
Prentice-Hall of India Private Limited, New Delhi
Prentice-Hall of Japan, Inc., Tokyo
Prentice-Hall (Singapore) Pte Ltd.
Editora Prentice-Hall do Brasil, Ltda., Rio De Janeiro

Contents

Foreword

Thirty years of careful study by physicists and others interested in education reveals that students enter our introductory courses with conceptual beliefs that differ considerably from the accepted concepts of physics. Unfortunately, a good fraction of these same students leave our courses with their alternative beliefs unchanged. Because of these studies, there is now more interest in helping students build a better conceptual foundation for their studies in physics.

Building this foundation is aided by uncovering students' alternative beliefs, a task well-suited for Ranking Tasks. A ranking task provides a question with several contextually similar situations. The situations differ in the value of one or more physical quantities (number of parallel or series bulbs, speed and mass of a swinging object, and so forth). The student ranks the situations according to some other physical quantity (electric current, rope tension and so forth). The student also provides reasons for their ranking. The ranking order and the explanation provide a window into the student's mind and helps a teacher or professor identify different models for what students are thinking.

Ranking tasks are also very useful for helping students modify their conceptual beliefs. Students are given a ranking task. The student completes the task working alone. The student then compares their thinking with that of another student and tries to reconcile any differences in thinking. The ranking task serves as the focus for a small group or classroom discussion. The tasks can also be given as homework problems and on quizzes and exams.

The present set of materials was developed during a series of workshops for two-year college physics professors which was supported by the National Science Foundation and organized and facilitated by Curt Hieggelke and Tom O'Kuma. David Maloney, the inventor of Ranking Tasks, was a co-leader for one series of these workshops. David taught participants how to use ranking tasks and then asked them to develop one or more tasks. Participants found the use and development of these tasks a valuable and enjoyable part of these intensive three-day workshops.

Do they make a difference? The gains of workshop participants' students on the Force Concept Inventory conceptual test and on the Mechanics Baseline problem-solving test have been excellent (see for example, TO-C and M-PD95b-C in Figure 4 on page 68 of Hake[1]). Curt Hieggelke's students, who use ranking tasks extensively, have some of the highest scores on the recently developed Conceptual Survey of Electricity and Magnetism. Perhaps more important is the ease with which faculty can learn to use and develop ranking tasks.

The authors (O'Kuma, Maloney and Hieggelke and the many talented two-year college professors who contributed to this book) have provided us with a rich resource of learning activities. Give them a try—you'll like them.

Alan Van Heuvelen

---

[1] R.R. Hake, "Interactive-engagement versus traditional methods: A six-thousand-student survey of mechanics test data for introductory physics courses," *Am. J. Phys.* **66**, 64-74 (1998).

Preface

On November 21–23, 1991, a workshop for two-year college physics teachers on Conceptual Exercises and Overview Case Studies was held at Joliet Junior College, Joliet, IL. This workshop was the first of many such workshops conducted as part of the Two-Year College (TYC) Physics Workshop Project, which was sponsored by Joliet Junior College, Lee College (Baytown, TX), and a series of grants from the Division of Undergraduate Education of the National Science Foundation. At one of the sessions in this workshop, the participants, working together in groups of three or four, constructed ranking tasks and then presented them to the larger group for feedback and response. (A ranking task is a conceptual exercise created by David P. Maloney as one of the many ways to ascertain a student's understanding of concepts.)

During the 1991–98 period, the TYC Physics Workshop Project continued to hold workshops during which a variety of ranking tasks for introductory physics were developed at or as part of the follow-up activities after the workshop by the participants. During the academic year 1993–94, the ranking tasks that had been developed as part of the TYC Physics Workshops were categorized and put into electronic form at Lee College. These were called *A Collection of Physics Ranking Tasks* and were distributed to previous workshop participants in May 1994. During the academic year 1994–95, a subset of these ranking tasks and newly developed ones were further refined and widely distributed by Joliet Junior College as *A Selection of Physics Ranking Tasks*. This book is a revision and expansion of that book.

Although these materials are copyrighted by Prentice Hall, professors have the right to use the materials for noncommercial educational purposes and can copy or have copied the materials in the book for the students in their classes. However, these materials, whether in their original or in an altered form, may not otherwise be distributed, transmitted, or included in other documents without express written permission from the publisher. We have included a CD with this book that has a pdf version of each ranking task exercise in 8 1/2 x 11 format to further facilitate professors' use of these materials in their own classes, workshops, or for any noncommercial use. If you do not already have Adobe Acrobat Reader, please contact Adobe's web site (http://www.adobe.com), where it can be downloaded for free. This software will allow you to print and/or modify these ranking tasks.

In the first section, we have included some background material, including sample and practice ranking tasks, which give students an understanding of this type of instrument, and examples of how to complete a ranking task. The main body of ranking tasks is divided into broad topical areas, with mechanics in the first section. The second section includes properties of matter, heat and thermodynamics, and waves. The third section covers electricity and magnetism. And finally, we have included an answer key for this edition.

You are invited to visit our web site for corrections, updates, and additional ranking tasks (http://www.tycphysics.org). If you have any questions about these ranking tasks, please contact us.

Tom O'Kuma   tokuma@lee.edu
Dave Maloney   maloney@ipfw.edu
Curtis Hieggelke   curth@jjc.cc.il.us

Acknowledgments

First, we want to acknowledge and thank all the participants who attended our workshops and were willing to share their ideas and work. Also, special thanks to all the workshop and conference leaders who made this project so successful.

Next, we want to give special thanks to the National Science Foundation for their support of the Two-Year College Physics Workshop Projects. We thank the many program directors at NSF that have supported this project, especially Duncan McBride, William Rauckhorst, William Kelly, Ken Krane, Ruth Howes, J. D. Garcia, Jack Hehn, and Karen Johnston.

At Joliet Junior College we particularly thank and acknowledge the support of this project by Michael Lee, Chairman of the Natural Science Department; Vice Presidents James Lepanto, Denis Wright, and J. D. Ross; and Presidents Raymond Pietak, Thomas Gamble, and J. D. Ross.

In addition, at Lee College we want to give special thanks to Wayne Miller and Carolyn Foster, Division Chairs of the Math, Science, and Engineering Division; Dean Donnetta Suchon; and Presidents Vivian Blevins and Jackson Sasser.

We want to thank Karen Dailey, who spent considerable time converting hard copy ranking tasks into electronic form at Lee College. We are also grateful for the assistance in various areas of this book by David Pequeno, Reggie Delgado, Regina Barrera, Sherry Green, Greg Gober, and Kathy O'Kuma.

At Joliet Junior College, we want to thank Craig Sumner, who reworked many diagrams, bringing them into final form. We also appreciate the help at different stages in this book by Jan Coleman, Judy Bond, Aaron Stone, Matt Kelly, Steve Clark, Eric Manuel, Geoff White, and Gale Ruggiero. Also, special thanks to physics professors Joe Krivicich (Joliet Junior College), William Hogan (Joliet Junior College), and Duane Desbain (Highland Community College, Highland, KS), who reviewed many of these ranking tasks.

We also want to thank our students, who tried out these ranking tasks and provided us with valuable feedback.

This material is based upon work supported by the Division of Undergraduate Education of the National Science Foundation under grant numbers USE 9150334, USE 9154271, DUE9255466, DUE 9353998, and DUE 9554683. Any opinions, findings, and conclusions or recommendations are those of the authors and do not necessarily reflect the views of the National Science Foundation.

Finally, we thank Alison M. Reeves at Prentice Hall, who supported this project and the publication of this book.

Contributors and Institutions

Albers, Dennis	Columbia College	Sonora	CA
Bennie-George, Toni	Green River Community College	Auburn	WA
Bryant, Light	South Mountain Community College	Phoenix	AZ
Cole, Jonathan	Mira Costa Community College	Oceanside	CA
D'Alessandris, Paul	Monroe Community College	Rochester	NY
Davies, Chad	Cloud County Community College	Concordia	KS
Dickison, Alex	Seminole Community College	Sanford	FL
Diesslin, Butch	Vermilion community College	Ely	MN
Eckard, Eldon	Bainbridge College	Bainbridge	GA
Emerson, Bruce	Central Oregon Community College	Bend	OR
Ezrailson,Cathy	Montgomery College	Conroe	TX
Fang, Chuen Chuen	Shelton State Community College	Tuscaloosa	AL
Fazzari, Greg	Walla Walla Community College	Walla Walla	WA
Golden, Peter	Isothermal Community College	Spindale	NC
Gundlach, Jim	John A. Logan College	Carterville	IL
Heath, Sally	Santa Rosa Community College	Santa Rosa	CA
Hehn, Jack	Tarrant County Junior College/AAPT	Hurst	TX
Hieggelke, Curtis	Joliet Junior College	Joliet	IL
Hill, Charlotte	Tarrant County Junior College	Hurst	TX
Johnson, Robert	Del Mar College	Corpus Christi	TX
Kaasa, Bruce	Iowa Central Community College	Fort Dodge	IA
Karmon, Oshri	Diablo Valley College	Pleasant Hill	CA
Keramati, Beta	Holmes Community College	Goodman	MS
Krupp, Robert	Wilber Wright College	Chicago	IL
Lam, Clement	North Harris College	Houston	TX
Loucks, Stuart	Solano Community College	Suisun	CA
Maloney, David	Indiana University–Purdue University Fort Wayne	Fort Wayne	IN
Mann, Myron	Los Angeles Valley College	Van Nuys	CA
Marquard, Paul	Casper College	Casper	WY
Nelson, Marvin	Green River Community College	Auburn	WA
Nicholson, Nick	Central Alabama Community College	Alexander City	AL
O'Kuma, Tom	Lee College	Baytown	TX
Pandey, Umesh	Albuquerque Technical–Vocational Institute	Albuquerque	NM
Plumb, Marie	Jamestown Community College	Jamestown	NY
Popp, John	Moraine Valley Community College	Palos Hills	IL
Saavrda, Sherry	Lake Sumter Community College	Leesburg	FL
Schramme, Dave	Rogers State College	Claremore	OK
Selgrad, Ot	Moraine Park Technical College	Fond du Lac	WI
Shepherd, Gordon	Guilford Technical Community College	Jamestown	NC
Speers, Bob	Fireland College – Bowling Green St. Univ.	Huron	OH
Splett, John	Erie Community College – City Campus	Buffalo	NY
Szalai, Imre	Delaware County Community College	Media	PA
Takahashi, Leo	Penn State – Beaver Campus	Monaca	PA
Taylor, Jack	Baltimore City Community College	Baltimore	MD
Thompson, Winston	Montgomery College	Takoma Park	MD
Ting, Cheng (David)	Southeast College–Houston Community College	Houston	TX
Van Heuvelen, Alan	The Ohio State University	Columbus	OH
Weiner, Adam	Green River Community College	Auburn	WA
West, Myra	Kent State University	Kent	OH
Wetz, Karen	Manatee Community College	Venice	FL
Willis, Bob	New Mexico Military Institute	Roswell	NM

Background, Insights, and Uses

This book is intended as a resource for physics instructors who are looking for tools to incorporate more conceptual analysis in their courses. In putting together this collection of ranking tasks (RTs), we have been guided by two major goals.

First, we wanted to provide instructors with a set of immediately useful RTs from as many topic domains in physics as possible. Although the number of RTs on different topics is not uniform, this collection does contain RTs from all topics of classical physics except for optics. (Optics has proven to be a topic for which it is difficult to design good conceptual ranking tasks.) The RTs in this manual can be copied directly from the manual and used immediately in the classroom. Or an instructor can quickly modify one of the items in the book to fit his or her style and context.

Second, we wanted to illustrate a wide variety of RTs so that instructors could get an idea of the flexibility of the format and ideas about how to generate RTs for themselves. Generation or examination of a ranking task will often suggest an easily produced variation or variations. Student responses to a ranking task are another good source of ideas for other RTs. We hope that seeing a range of variations for a variety of RTs will enable instructors to develop sufficient familiarity with RTs so that they will be able to develop their own. Constructing good ranking tasks is not easy, but interested instructors should be able to generate viable ones with some practice.

Nature and Structure

The idea of ranking tasks arose from research into students' conceptions using a technique called rule assessment, developed by Robert S. Siegler (1976). The rule-assessment technique involved having subjects make a comparative judgment about a large variety of arrangements of a specific situation. The ranking tasks were conceived as a shorter format for eliciting such comparative judgments.

A ranking task is a paper-and-pencil exercise that presents students with a set of variations on a particular physical situation. The students are supposed to rank the variations on a specified basis. After explicitly writing out their ranking sequence or choosing the option that all of the variations are equivalent, the students are asked to write out an explanation of their reasoning. Finally, students are given an opportunity to identify how sure they were of the reasoning they used in the task.

As explained in the original ranking tasks article (Maloney, 1987) the basic structure of RT has four elements: the description of the situation, including the constraints and the basis for ranking the arrangements; a set of figures showing the different arrangements to be compared; a place to identify the response sequence chosen or to indicate that all of the arrangements have the same value for the ranking basis; and a place to explain the reasoning for the answer produced. Many of the RTs in this book have an additional element–a scale at the bottom of the page for the students to indicate how sure they were about their answers. This can be useful in at least two ways. First, it can tell an instructor how strongly the students are consciously committed to their ideas. Second, many students often feel that they are "just guessing" when they answer the RT, and that is what they want to write for their explanation of the reasoning. Having the scale at the bottom of the page provides an outlet for the students to express these feelings so they can use the explanation section to describe what they actually did when working the task.

Ranking tasks contain few clues about how they should be worked. In addition, they require students to think about the situations in an unusual manner. Usually students deal with such situations in physics by calculating a specific numerical value given a limited number of other numerical values, all of which are normally needed in the calculation. In a ranking task students are confronted with a set of variations that usually differ in the specific values for two variables, and they have to decide how these variables affect the behavior of interest of the system.

Although most of the ranking tasks in this manual contain numerical values for two variables, we think of RTs as conceptual exercises. One might wonder how this works. The reason for our contention is that experience has shown that students often use the numerical values in inappropriate ways. Such use reveals one of several problems. One common problem is that the students do not understand the relations they are using, but rather just know to plug whatever numerical values are available into whatever relation is available. A second common problem is for students to apply the wrong relation to the situation. An example of this difficulty is students' strong tendency to use the product of mass and speed when they actually need to use the kinetic energy. So having the students use the numerical values in inappropriate ways can provide insight into their concepts and strategies.

The RTs in this manual vary widely in a number of ways. One way in which they vary is in how much computation is required to do a task. There are several reasons for this variation. One case is the "Two Different Blocks and a Pulley–Net Force" task on page 30, where students need to set up Newton's second law for the two masses and solve the equations simultaneously. Students seldom approach this task in this manner, but they also lack an equation they can quickly apply, so presenting them with such a situation makes them generate a procedure. It is often useful to see what students do with such a task. Sometimes we do want the students to learn how to do certain calculations, e.g., "Five T's Rotating About an Axis (top view)–Angular Acceleration" on page 94. And in some cases students need to do the calculations to understand that apriori ideas (e.g., symmetry considerations) don't always work out.

Reasons for Using Ranking Tasks

A fairly obvious question at this point is why an instructor would want to use ranking tasks. One strong reason for using them is the fact that they frequently elicit students' natural ideas about the behavior of physical systems rather than a memorized response. This ability of the ranking tasks to elicit students' natural ideas provides instructors with a way to gain important insights into students' thinking. With the help of those insights the instructor can help the students adapt to the scientifically accepted ideas.

Research in physics education (Clement, 1982; Peters, 1982; Halloun and Hestenes 1985; McDermott, 1991) has demonstrated that it is often difficult to get students to change their natural ideas about the physical world. A productive part of any effort to change students' ideas can be getting them to consider the same idea in a variety of ways. In a sense we want to be able to ask the same question in a variety of ways. Ranking tasks are a useful tool in such an effort, since they provide a way to frame questions in a manner that is novel for almost all students. Subsequently showing the students that they have responded in a different way to the same question asked in different ways can be useful for getting them to think about why they responded as they did in each case. That, in turn, can lead them to think about which, if either, response they believe in more strongly and why.

Another important reason for using RTs is that they can be used to reframe a question asked in a traditional problem, or a multiple choice item, or an essay, but in a different way. So RTs can be used to determine how robust a student's concept is. Students get used to responding to multiple choice items, and they often develop coping mechanisms that allow them to respond to such items without really thinking the situation through. Couching the same question in a different format, which requires a different way of evaluating the situation, makes the students think about the concepts, principles, and relations in another way.

We have found that certain RTs are excellent for helping students develop legitimate understanding of some concepts. For example, there are a number of ranking tasks in this book involving uniform electric fields. It has been our experience that if students are not asked about uniform electric fields, they have definite misconceptions about them. However, after some work with ranking tasks, such as those in this book, most students develop a robust understanding of this issue. We believe that there are a number of ideas for which well-designed ranking tasks are an especially effective way to help students learn.

Uses

Ranking tasks are useful in a variety of ways. They make good homework assignments and good test questions. Ranking tasks are a good "size" and form as homework assignments because they are simple and easy for the students to understand but require careful and thorough analysis for correct completion. But RTs can be made challenging enough that they can even be assigned as homework where the students are allowed (encouraged) to work together. An example of such an item is "Circuit with Three Open and Closed Switches–Voltmeter Readings I" on page 186.

As test items RTs provide two parts–the ranking sequence and the explanation–that can be scored separately; for example, 2 points of 5 for the correct ranking sequence and 3 of 5 for a correct explanation. RTs are usually challenging test items for students. When using RTs as test items it is often best to have only four or five variations so that the students do not spend too much time, and/or make silly errors, with the calculations.

Ranking tasks are also very useful if an instructor wants to generate class discussions or have students engage in peer instruction or group work. A productive way to generate a class discussion is to give the class a ranking task, allow them about 5 minutes to work it, have them talk to each other about their solutions, and then ask selected individuals to present their answers. The instructor can then either have these students defend their ideas, assuming conflicting ideas have been presented, or can talk about the ideas and how to reason to the correct response. Giving ranking tasks to the students, who have been assigned to small groups, and telling each group that they have to come to a consensus about the answer is another good way to use RTs.

Sets of related tasks are also good for class discussions. Tasks can be related by having different objects exhibiting the same physical behavior, e.g., an arrow, a stream of water, and a rock undergoing projectile motion, by changing the variables for the situation, e.g., mass and speed, mass and height, or speed and height for a ball thrown off a building, or by a number of other modifications of a situation. After the students work individual RTs from the related set, the students can be grouped by the particular version of the task they worked to discuss their answers. (This assumes the class is small enough for them to move around the room.)

Ranking tasks can be used to pre- and posttest students to check on the extent to which their ideas have been changed by instruction. Since RTs often bypass students' memorized physics ideas, they are especially useful in determining the extent to which instructors have been successful in changing students' natural (common sense) beliefs about physical situations.

The first time students are presented with ranking tasks, we suggest first giving them one of the sample ranking task pages. Give them an opportunity to read through the page carefully and check to see if they have any questions about how to work these tasks. Then we suggest passing out the practice RT and having them work it. Go over the answer, emphasizing explicitly showing how ties in the answer sequence are indicated and writing a complete explanation.

There is one very important point about using RTs in situations where students will earn credit for doing them. Since RTs are unfamiliar to students, it is critical for the students to have an opportunity to practice with the format in a noncredit context first. This is important to assure that students explicitly show ties and that they write complete explanations.

Related Ranking Tasks

If an instructor decides that they want to generate their own ranking tasks, a good way to start is to generate a variation on an existing task. There are several ways to vary existing tasks to produce new tasks.

One way is to have students rank the same situation on a different basis. An example of this technique is found in the "Cars and Barriers–Stopping Distance/Time with the Same Force" tasks on pages 62 and 80. In a similar way it is often possible to ask the same question in different ways. An example of this is "Carts Moving Along Horizontal Surface" on pages 17 and 18. This approach is especially useful when one version of the question uses the technical language of physics while the other employs natural, everyday language. Another related technique is to ask the same question but have different variables for the students to work with.

A different approach is to have the same variables and question but vary the physical situation. An example of this is shown in "Horizontal Arrows," "Rifle Shots," "Toy Trucks," and "Spheres Thrown Horizontally Off Cliffs" tasks on pages 48 – 51. All of these are the same projectile motion situation, however, students often think that different physical systems will behave differently. Getting students to see past these noncritical surface features to the physical principle that applies is an important aspect of learning to do physics.

Another very important aspect that can be varied is the representation used for presenting the information. An example of this is shown in "Uniform Electric Field—Electric Force on Charge at Rest" tasks on pages 150–152. Students often learn information in one representation, i.e., they will know what to do if given a kinematic equation, but will be lost if the same situation and information is presented in graphical form. Ability to handle tasks represented in various ways is a good indicator of solid understanding.

It will be easier to construct variations for some situations, concepts, or principles than for others. And for some situations, concepts, or principles students may need to work through a wider variety of those variations in order to develop a solid understanding. Our experience indicates that the topic of sinking, floating, and buoyancy is one topic where a wider variety of tasks is useful. Consequently, both to present ideas about generating related RTs and to give a specific example of how to approach such a topic, there is what will probably seem to many a large number of RTs on sinking, floating, and buoyancy.

A unique contribution to this book is the Resistive Circuits Concepts Diagnostic Test developed by Dennis Albers. This test is a sequence of ranking tasks on basic electric circuit concepts. In a way this test takes the idea of using related ranking tasks to the extreme, but having this set of interrelated ranking tasks enables an instructor to get a clear understanding of students' ideas about simple electric circuits. Similar tests could obviously be developed for other physics topics.

Finally some other ideas for different types of RTs, such as using comparative values for variables rather than specific numerical values and other ways to develop tasks that involve three variables, can be found in Maloney and Friedel (1996).

In writing RTs we believe there are several issues that should be carefully considered. First, we believe instructors should take account of known, or suspected, student alternative conceptions when developing an RT. Because RTs frequently elicit students' natural ideas, using those ideas in writing the RT can make it a better conceptual exercise. A second issue is the language—natural, i.e., everyday, versus technical—used in the RT. If one wants to elicit students' common-sense ideas, then natural language is usually more effective. But having students respond to the same task written in natural and technical language can be a useful exercise. A third issue is how many variations to present in the RT. In some situations this can be fixed by the situation, which may have a fixed number of variations. In most other cases the number of variations to include is usually tied to how the RT is to be used. As a homework assignment, eight variations might be chosen to provide practice in calculating and evaluating alternatives. For a test item, between four and six variations are more reasonable.

The editors would appreciate feedback about the items in this book. We are especially interested in learning about other variations and uses of ranking tasks and about new RTs. We would also like to hear about any errors that may have crept in despite our best efforts to ensure that everything is correct.

Clement, J. (1982) Students' Preconceptions in Introductory Mechanics, *American Journal of Physics* **50**: 66-71.

Halloun, I. A. and Hestenes, D. (1985) The Initial Knowledge State of College Physics Students, *American Journal of Physics* **53**: 1043-1055

McDermott, L. C. (1991) Millikan Lecture 1990: What we Teach and What is Learned-Closing the Gap, *American Journal of Physics* **59**: 301-315.

Maloney, D. P. (1987) Ranking Tasks; A New Type of Test Item, *Journal of College Science Teaching* **16**(6): 510.

Maloney, D. P. and Friedel, A.W. (1996) Ranking Tasks Revisited , *Journal of College Science Teaching* **25**: 205-210.

Peters, P. C. (1982) Even Honors Students Have Conceptual Difficulties with Physics, *American Journal of Physics* **50**: 501-508.

Siegler, R. S. (1976) Three Aspects of Cognitive Development, *Cognitive Psychology* **8**: 481-520.

Ranking Task Sample I

For a ranking task, each item will have a number of situations as illustrated. Your task will be to rank the items in a specific order. After ranking them you will be asked to identify the basis you used for the ranking and the reasoning behind your choice. It is extremely important that you are careful to write out the proper ranking once you have determined what basis you are going to use, i.e., make sure all of the situations are ranked in the proper order according to your basis. The sample below shows how to rank items and what your explanation should be like. **NOTE: Although the procedure for working the item is correct, the particular answer, which was chosen at random from actual student responses, may not be correct.**

Example:

Shown below are eight cars that are moving along horizontal roads at specified speeds. Also given are the masses of the cars. All of the cars are the same size and shape, but they are carrying loads with different masses. All of these cars are going to be stopped by plowing into barrel barriers. All of the cars are going to be stopped in the same distance.

Rank these situations from greatest to least on the basis of the strength of the forces that will be needed to stop the cars in the same distance. That is, put first the car on which the strongest force will have to be applied to stop it in x meters, and put last the car on which the weakest force will be applied to stop the car in the same distance.

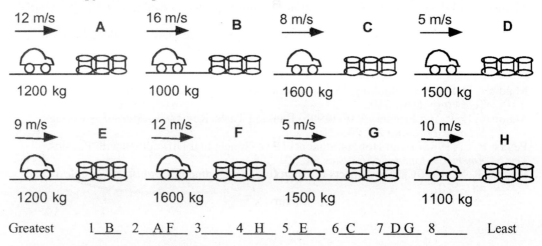

Greatest   1 _B_ 2 _A F_ 3 ____ 4 _H_ 5 _E_ 6 _C_ 7 _D G_ 8 ____   Least

Or, all cars require the same force. _____

Please carefully explain your reasoning

Since acceleration is the change in velocity divided by the change in time and all the changes in times are the same, then I used the change in velocity.

How sure were you of your ranking? (circle one)
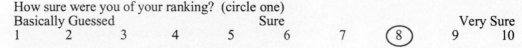

Basically Guessed   Sure   Very Sure
1   2   3   4   5   6   7   (8)   9   10

Notice in this example that two situations produced the same result for the ranking and that these were listed in the same answer blank. Such a possibility exists for all items. In the same way, it is possible that all of the situations will give the same result. If that occurs, and only if that occurs, the option of all equal, or all the same, should be chosen.

Ranking Task Sample II

Each ranking task will have a number of situations, or variations of a situation, that have varying values for two or three variables. Your task is to rank these variations on a specified basis. After ranking the items, you will be asked to explain how you determined your ranking sequence and the reasoning behind the way you used the values of the variables to reach your answer. An example of how to work the ranking tasks follows.

Example:

Shown below are six situations where a cart, which is initially moving to the right, has a force applied to it such that the force will cause the cart to come to a stop. All of the carts have the same initial speed, but the masses of the carts vary, as do the forces acting upon them.

Rank these situations, from greatest to least, on the basis of how long it will take each cart to stop.

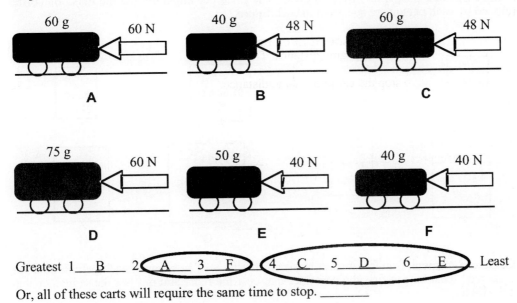

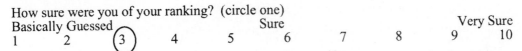

Greatest 1 __B__ 2 __A__ 3 __F__ 4 __C__ 5 __D__ 6 __E__ Least

Or, all of these carts will require the same time to stop. _____

Please carefully explain your reasoning.

I think the time depends on the acceleration, so I divided the forces by the masses.

How sure were you of your ranking? (circle one)
Basically Guessed   Sure   Very Sure
1   2   ③   4   5   6   7   8   9   10

Notice in this example that in one instance, two of the situations produced the same value of the ratio used to determine the ranking, and that the letters for the ones that tied are circled showing they were ranked equally (A and F). In another instance, three of the remaining situations have the same ranking and they are circled together (C and D and E), showing this result. In the same way, it is possible that all of the arrangements will give the same result for a particular basis. If that occurs, and only if that occurs, the option of all equal, or all the same, should be chosen.

Practice Ranking Task

The purpose of this page is to practice an exercise known as a ranking task. Each ranking task will have a number of situations, or a number of variations of a situation, that have varying values for two, three, or four variables. Your task is to rank these variations using the basis specified in the problem statement. After ranking the items, you will be asked to explain how you determined your ranking sequence and the reasoning behind the way you worked the item as you did. In addition, you should indicate your confidence in your ranking and reasoning. Notice that *it is very important to show such ties explicitly*! It is possible that all of the variations could have the same value on the ranked basis. If that occurs, *and only if that occurs*, you should choose the all equal, or all same, option.

Example:

Shown below are eight rectangles representing containers of coffee. These containers were purchased by eight people at various stores. The price per kilogram and the mass of coffee purchased by each person are specified in each figure.

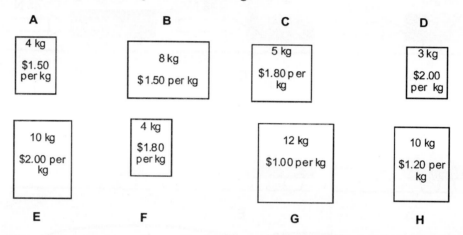

Rank these containers, from greatest to least, on the basis of how much each person paid for their coffee. That is, put first the container that cost the most and put last the container that cost the least.

Greatest   1_____ 2_____ 3_____ 4_____ 5_____ 6_____ 7_____ 8_____   Least

Or, the cost was the same for all of these containers _____

Please carefully explain your reasoning:

How sure were you of your ranking? (circle one)
Basically Guessed   Sure   Very Sure
   1   2   3   4   5   6   7   8   9   10

RANKING TASK
EXERCISES IN PHYSICS

Kinematics Ranking Tasks

Ball Motion Diagrams—Velocity I [1]

The following drawings indicate the motion of a ball subject to one or more forces on various surfaces from <u>left to right</u>. Each circle represents the position of the ball at succeeding instants of time. Each time-interval between successive positions is equal.

Rank each case from the highest to the lowest velocity based on the ball's last velocity using the coordinate system specified by the dashed arrows in the figures. Note: Zero is greater than negative, and ties are possible.

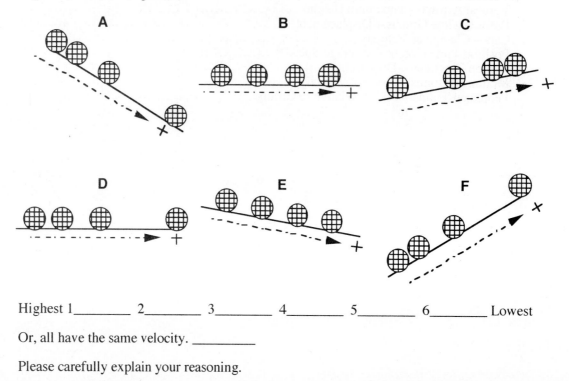

Highest 1_____ 2_____ 3_____ 4_____ 5_____ 6_____ Lowest

Or, all have the same velocity. _____

Please carefully explain your reasoning.

How sure were you of your ranking? (circle one)
Basically Guessed   Sure   Very Sure
1   2   3   4   5   6   7   8   9   10

[1] D. Schramme, C. Fang, B. Speers
Physics Ranking Tasks

Mechanics

Ball Motion Diagrams—Acceleration I [2]

The following drawings indicate the motion of a ball subject to one or more forces on various surfaces from left to right. Each circle represents the position of the ball at succeeding instants of time. Each time-interval between successive positions is equal.

Rank each case from the highest to the lowest acceleration, based on the drawings. Assume all accelerations are constant and use the coordinate system specified in the drawing. Note: Zero is greater than negative acceleration, and ties are possible.

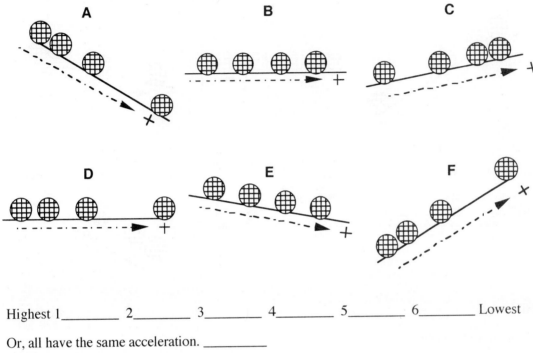

Highest 1_____ 2_____ 3_____ 4_____ 5_____ 6_____ Lowest

Or, all have the same acceleration. _____

Please carefully explain your reasoning.

How sure were you of your ranking? (circle one)
Basically Guessed   Sure   Very Sure
1   2   3   4   5   6   7   8   9   10

[2] D. Schramme, C. Fang, B. Speers
Physics Ranking Tasks

Mechanics

The following drawings indicate the motion of a ball subject to one or more forces on various surfaces from <u>left to right</u>. Each circle represents the position of the ball at succeeding instants of time. Each time-interval between successive positions is equal.

Rank each case from the highest to the lowest velocity based on the ball's last velocity using the coordinate system specified in the drawing. Note: Zero is greater than negative, and ties are possible.

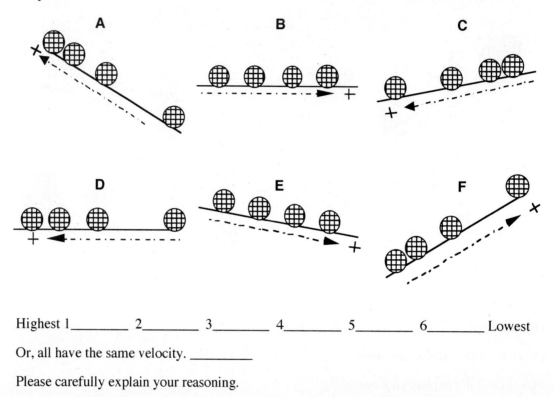

Highest 1_____ 2_____ 3_____ 4_____ 5_____ 6_____ Lowest

Or, all have the same velocity. _____

Please carefully explain your reasoning.

How sure were you of your ranking? (circle one)
Basically Guessed   Sure   Very Sure
1   2   3   4   5   6   7   8   9   10

[3] D. Schramme, C. Fang, B. Speers, C. Hieggelke, D. Maloney, T. O'Kuma

The following drawings indicate the motion of a ball subject to one or more forces on various surfaces from <u>left to right</u>. Each circle represents the position of the ball at succeeding instants of time. Each time-interval between successive positions is equal.

Rank each case from the highest to the lowest acceleration, based on the drawings. Assume all accelerations are constant and use the coordinate system specified in the drawing. Note: Zero is greater than negative acceleration, and ties are possible.

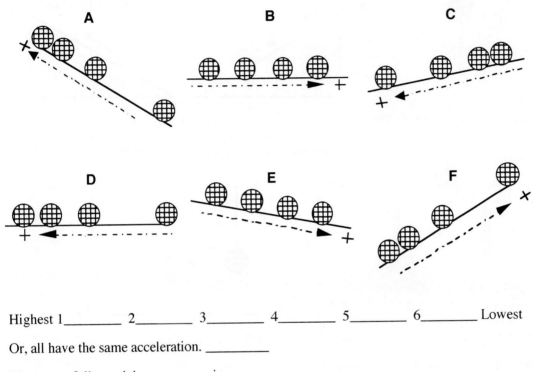

Highest 1_____ 2_____ 3_____ 4_____ 5_____ 6_____ Lowest

Or, all have the same acceleration. _____

Please carefully explain your reasoning.

How sure were you of your ranking? (circle one)
Basically Guessed   Sure   Very Sure
1   2   3   4   5   6   7   8   9   10

[4] D. Schramme, C. Fang, B. Speers, C. Hieggelke, D. Maloney, T. O'Kuma

Objects in Different Situations—Accelerations [5]

The following objects are sitting, falling, rolling, swinging, or just going in circles, as indicated in the different situations below. Each has an acceleration in some direction or another. You are to rank them from greatest to least on the basis of the magnitude of acceleration. If two or more objects have the same acceleration, rank them the same.

A. Object dropped from the top of a building.

B. Object rolling from rest down an inclined plane.

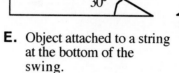

C. Object rolling up an inclined plane after being given an initial velocity of 4 m/s.

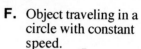

D. Object sitting on a table top motionless.

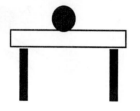

E. Object attached to a string at the bottom of the swing.

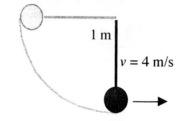

F. Object traveling in a circle with constant speed.

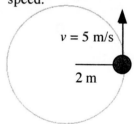

Greatest   1_____   2_____   3_____   4_____   5_____   6_____   Least

Or, all of these have the same magnitude of acceleration._____

Please carefully explain your reasoning.

How sure were you of your ranking? (circle one)
Basically Guessed   Sure   Very Sure
1   2   3   4   5   6   7   8   9   10

[5] K. W. Nicholson

Vertical Model Rockets—Maximum Height [6]

The eight figures below depict eight model rockets that have just had their engines turned off. All of the rockets are aimed straight up, but their speeds differ. All of the rockets are the same size and shape, but they carry different loads, so their masses differ. The specific mass and speed for each rocket is given in each figure. (In this situation, we are going to ignore any effect air resistance may have on the rockets.) At the instant when the engines are turned off, the rockets are all at the same height.

Rank these model rockets, from greatest to least, on the basis of the maximum height they will reach.

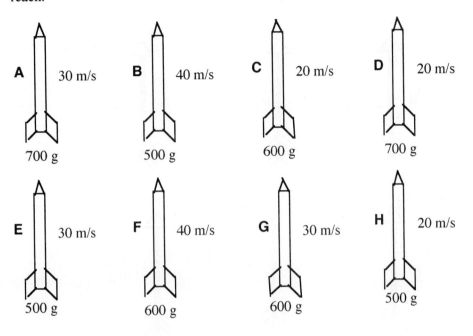

A 30 m/s 700 g   **B** 40 m/s 500 g   **C** 20 m/s 600 g   **D** 20 m/s 700 g

E 30 m/s 500 g   **F** 40 m/s 600 g   **G** 30 m/s 600 g   **H** 20 m/s 500 g

Highest 1_____ 2_____ 3_____ 4_____ 5_____ 6_____ 7_____ 8_____ Lowest

Or, all rockets reach the same height. _____

Please carefully explain your reasoning.

How sure were you of your ranking? (circle one)
Basically Guessed   Sure   Very Sure
   1   2   3   4   5   6   7   8   9   10

The eight figures below show arrows that have been shot into the air. All of the arrows were shot straight up and are the same size and shape. But the arrows are made of different materials so they have different masses, and they have different speeds as they leave the bows. The values for each arrow are given in the figures. (We assume for this situation that the effect of air resistance can be neglected.) All start from same height.

Rank these arrows, from greatest to least, on the basis of the maximum heights the arrows reach.

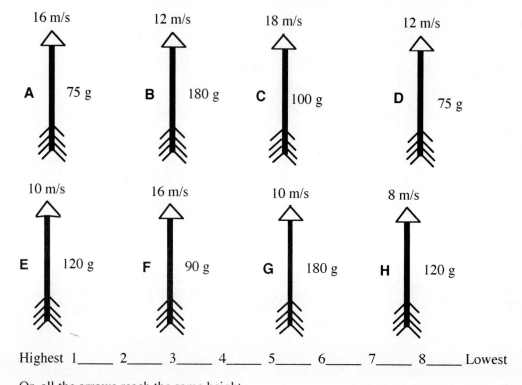

Highest 1_____ 2_____ 3_____ 4_____ 5_____ 6_____ 7_____ 8_____ Lowest

Or, all the arrows reach the same height. _____

Please carefully explain your reasoning.

How sure were you of your ranking? (circle one)

Basically Guessed   Sure   Very Sure

1   2   3   4   5   6   7   8   9   10

[7] T. O'Kuma

Position Time Graphs—Displacement [8]

In the position vs. time graphs below, all the times are in seconds (s), and all the positions are in meters (m). Rank these graphs on the basis of which graph indicates the greatest displacement from beginning to end of motion. Give the highest rank to the one(s) with the greatest displacement, and give the lowest rank to the one(s) indicating the least displacement. If two graphs indicate the same displacement, give them the same rank. Note: Zero is greater than negative, and ties are possible.

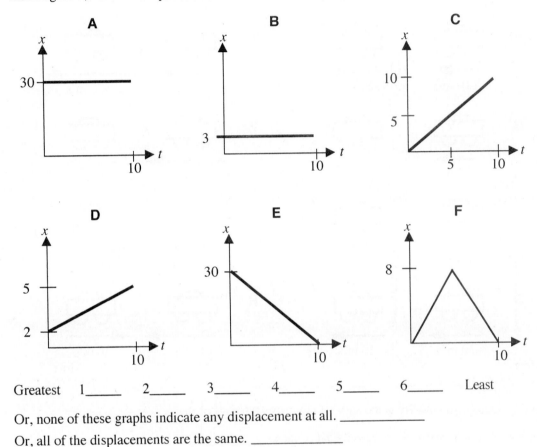

Greatest   1_____   2_____   3_____   4_____   5_____   6_____   Least

Or, none of these graphs indicate any displacement at all. _____

Or, all of the displacements are the same. _____

Please carefully explain your reasoning.

How sure were you of your ranking? (Circle one)
Basically Guessed   Sure   Very Sure
1   2   3   4   5   6   7   8   9   10

The eight situations below show *before* and *after* "snapshots" of a car's velocity. Rank these situations, in terms of the change in velocity, from most positive to most negative. All cars have the same mass and they traveled the same distance. Negative numbers, if any, rank lower than positive ones (-20 m/s < -10 m/s < 0 < 5).

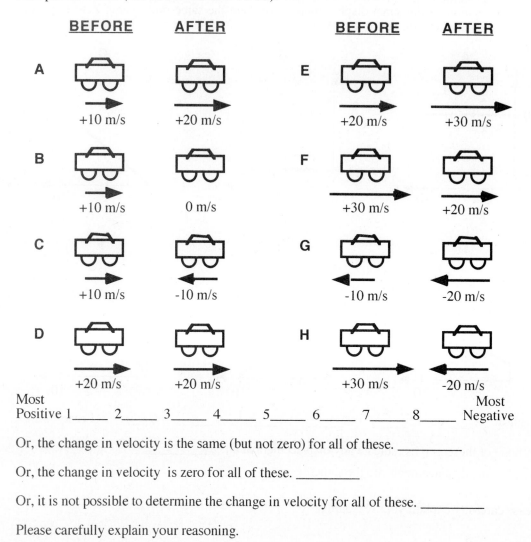

Most
Positive 1_____ 2_____ 3_____ 4_____ 5_____ 6_____ 7_____ 8_____ Most Negative

Or, the change in velocity is the same (but not zero) for all of these. _____

Or, the change in velocity is zero for all of these. _____

Or, it is not possible to determine the change in velocity for all of these. _____

Please carefully explain your reasoning.

How sure were you of your ranking? (circle one)

Basically Guessed					Sure				Very Sure
1	2	3	4	5	6	7	8	9	10

[9] J. Cole, D. Maloney, C. Hieggelke

Position Time Graphs—Average Speed [10]

In the position vs. time graphs below, all the times are in seconds (s), and all the positions are in meters (m). Rank these graphs on the basis of which graph indicates the greatest average speed, where the average speed is calculated from the beginning to the end of motion. Give the highest rank to the one(s) with the greatest average speed, and give the lowest rank to the one(s) indicating the least average speed. If two graphs indicate the same average speed, give them the same rank.

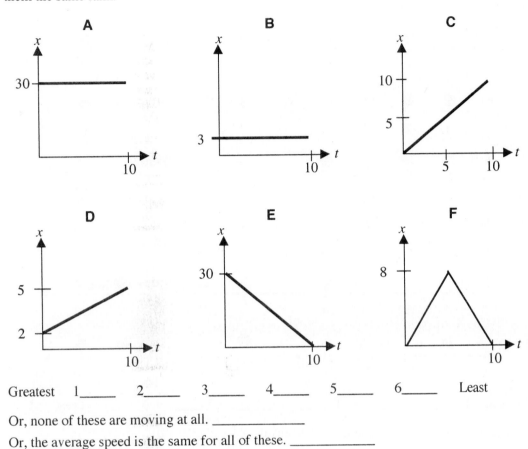

Greatest   1_____   2_____   3_____   4_____   5_____   6_____   Least

Or, none of these are moving at all. _____

Or, the average speed is the same for all of these. _____

Please carefully explain your reasoning.

How sure were you of your ranking? (circle one)

Basically Guessed				Sure				Very Sure	
1	2	3	4	5	6	7	8	9	10

Flash strobe photographs were taken every second of a set of spheres moving from left to right. The diagram below shows the location of each sphere when each photograph was taken. The total time intervals shown vary among the spheres. All the displacements are in meters. Rank these spheres on the basis of the greatest displacement over the first 3 seconds. Give the highest rank to the one(s) with the greatest displacement, and give the lowest rank to the one(s) indicating the least displacement. If two motion diagrams indicate the same displacement for the 3-second interval, give them the same rank.

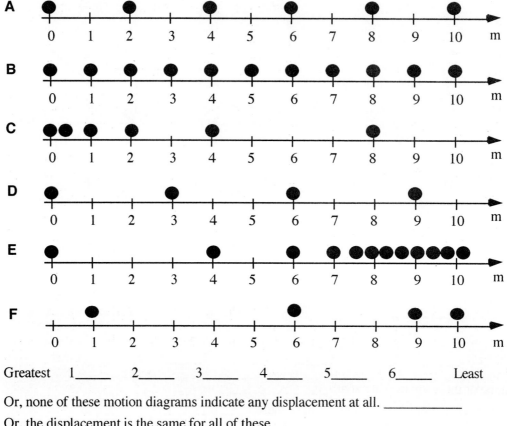

Greatest   1_____   2_____   3_____   4_____   5_____   6_____   Least

Or, none of these motion diagrams indicate any displacement at all. _____

Or, the displacement is the same for all of these. _____

Please carefully explain your reasoning.

How sure were you of your ranking? (circle one)
Basically Guessed   Sure   Very Sure
     1   2   3   4   5   6   7   8   9   10

[11] K. W. Nicholson, C. Hieggelke, D. Maloney

Motion Diagrams—Average Velocity [12]

Flash strobe photographs were taken every second of a set of spheres moving from left to right. The diagram below shows the location of each sphere when each photograph was taken. The total time intervals shown vary among the spheres. All the displacements are in meters. Rank these spheres on the basis of the greatest average velocity over the first 3 seconds. Give the highest rank to the one(s) with the greatest average velocity, and give the lowest rank to the one(s) indicating the least average velocity. If two diagrams indicate the same average velocity for the 3-second interval, give them the same rank.

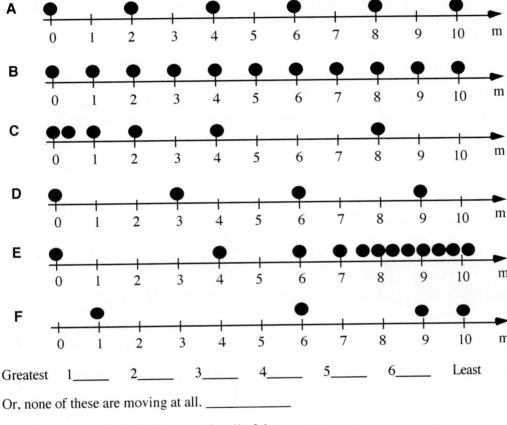

Greatest   1_____   2_____   3_____   4_____   5_____   6_____   Least

Or, none of these are moving at all. _____

Or, the average velocity is the same for all of these. _____

Please carefully explain your reasoning.

How sure were you of your ranking? (circle one)

Basically Guessed   Sure   Very Sure

1   2   3   4   5   6   7   8   9   10

[12] K. W. Nicholson, C. Hieggelke, D. Maloney

Shown below are six situations where people are walking or running through train cars. The people are walking/running at various speeds either in the direction the train is traveling or opposite to the direction of travel of the train. The speed of the train and the speed and direction of each person are given in each figure. An observer is standing beside the track watching the train go by.

Rank these walkers/runners, from greatest to least, on the basis of how fast they are moving relative to the observer standing beside the tracks. That is, put first the person the observer would say is going fastest and put last the person the observer would say was the slowest.

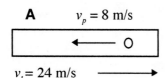

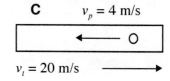

C $v_p = 4$ m/s

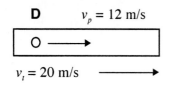

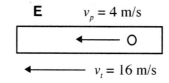

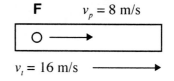

Greatest 1_____ 2_____ 3_____ 4_____ 5_____ 6_____ Least

Or, all six of these people have the same speed according to the observer. _____

Please carefully explain your reasoning.

How sure were you of your ranking? (circle one)
Basically Guessed   Sure   Very Sure
1   2   3   4   5   6   7   8   9   10

[13] C. Hieggelke, D. Maloney, T. O'Kuma

Force Ranking Tasks

Carts Moving Along Horizontal Surface—String Tension [14]

The six figures below show carts that are moving along horizontal surfaces at various speeds. The carts are the same size and shape but carry different loads, so their masses differ. All of the carts have a massless string attached, which passes over a frictionless massless pulley and is tied to a metal block that is hanging free. All of the metal blocks are identical. As the carts move to the right they pull the blocks up toward the horizontal surface, which is the top of the table.

Rank these situations, from greatest to least, on the basis of the tension in the strings at the instant shown. That is, put first the situation where the string is under the greatest tension, and put last the situation where the string is under the least tension at that instant.

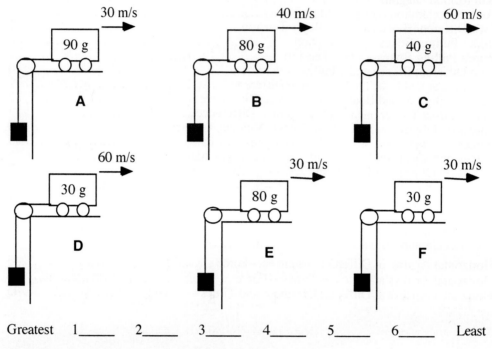

Greatest   1_____   2_____   3_____   4_____   5_____   6_____   Least

Or, all of these strings are under the same tension. _____

Or, there is no tension in any of these strings. _____

Please carefully explain your reasoning.

How sure were you of your ranking? (circle one)
Basically Guessed   Sure   Very Sure
1   2   3   4   5   6   7   8   9   10

[14] D. Maloney

The six figures below show carts that are moving along horizontal surfaces at various speeds. The carts are the same size and shape but carry different loads, so their masses differ. All of the carts have a string attached, which passes over a pulley and is tied to a metal block that is hanging free. All of the metal blocks are identical. As the carts move to the right, they will pull the blocks up toward the horizontal surface, which is the top of the table.

Rank these situations, from greatest to least, on the basis of the magnitude of the acceleration of the carts. That is, put first the situation where the cart has the greatest acceleration, and put last the situation where the cart has the smallest acceleration.

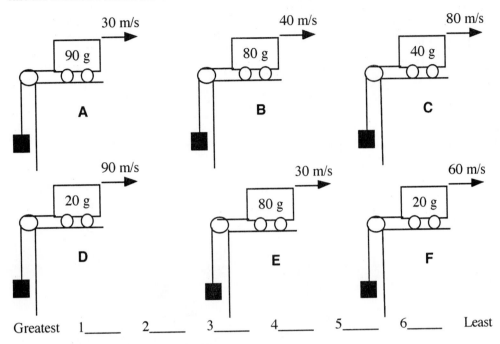

Greatest   1_____   2_____   3_____   4_____   5_____   6_____   Least

Or, all of these carts have the same magnitude acceleration. _____

Or, there is no acceleration in any of these carts. _____

Please carefully explain your reasoning.

How sure were you of your ranking? (circle one)
Basically Guessed   Sure   Very Sure
1   2   3   4   5   6   7   8   9   10

Carts Moving Along Horizontal Surface—Slowing Down [16]

The six figures below show carts that are moving along horizontal surfaces at various speeds. The carts are the same size and shape but carry different loads so their masses differ. All of the carts have a string attached, which passes over a pulley and is tied to a metal block that is hanging free. All of the metal blocks are identical. As the carts move to the right they pull the blocks up toward the horizontal surface, which is the top of the table.

Rank these situations, from greatest to least, on the basis of which cart is slowing down most quickly. That is, put first the situation where the cart is slowing down the quickest, and put last the situation where the cart is slowing down at the slowest rate.

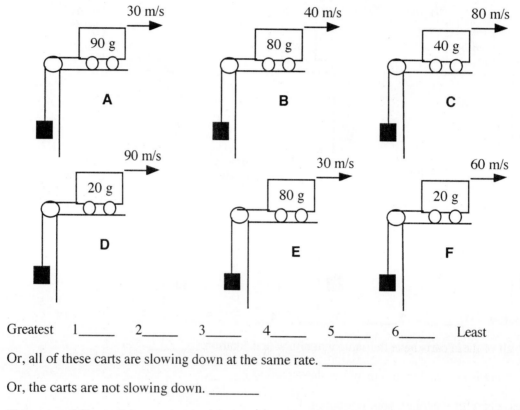

Greatest   1_____   2_____   3_____   4_____   5_____   6_____   Least

Or, all of these carts are slowing down at the same rate. _____

Or, the carts are not slowing down. _____

Please carefully explain your reasoning.

How sure were you of your ranking? (circle one)

Basically Guessed				Sure				Very Sure	
1	2	3	4	5	6	7	8	9	10

Two-Dimensional Forces on a Treasure Chest—Final Speed [17]

The six figures below show treasure chests with two forces acting upon them. The lengths of the force vectors represent the magnitudes of the forces. Rank these situations from greatest to least with regard to the final speed of the treasure chest after 2 seconds. All chests start at rest. If you believe that two of the situations have the same final speed, place both of their letters at the same rank.

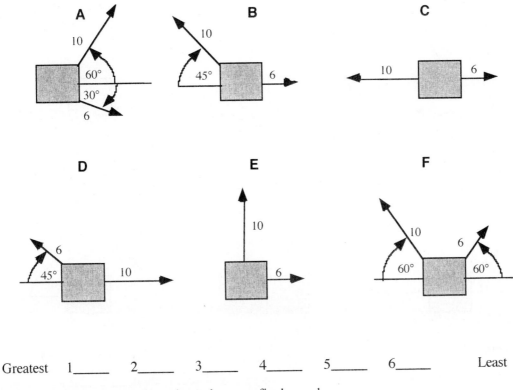

Greatest   1_____   2_____   3_____   4_____   5_____   6_____   Least

Or, all of these treasure chests have the same final speed. _____
Please carefully explain your reasoning.

How sure were you of your ranking? (circle one)
Basically Guessed   Sure   Very Sure
1   2   3   4   5   6   7   8   9   10

[17] K. W. Nicholson, D. Maloney, T. O'Kuma

The six figures below represent treasure chests with two forces acting on them. The lengths of the force vectors represent the magnitude of the force. Rank these situations from greatest to least with regard to the resulting magnitude of acceleration of the treasure chest. If you believe two situations have the same magnitude of acceleration, place both their letters at the same rank.

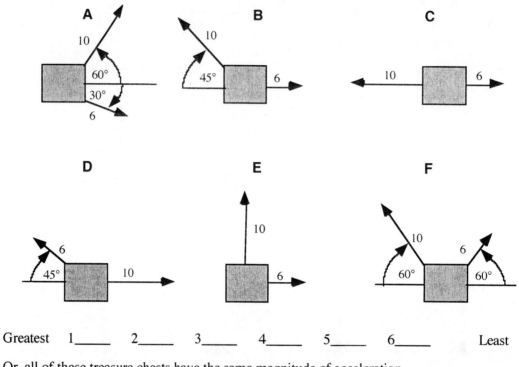

Greatest   1_____   2_____   3_____   4_____   5_____   6_____   Least

Or, all of these treasure chests have the same magnitude of acceleration. _____

Please carefully explain your reasoning.

How sure were you of your ranking?
Basically Guessed   Sure   Very Sure
1   2   3   4   5   6   7   8   9   10

[18] K.W. Nicholson, D. Maloney, T. O'Kuma

Arrows—Acceleration[19]

The eight figures below show arrows that have been shot into the air. All of the arrows were shot straight up and are the same size and shape. The arrows are made of different materials so they have different masses, and they have different speeds as they leave the bows. The values for each arrow are given in the figures. (We assume for this situation that the effect of air resistance can be neglected.) All start from same height.

Rank these arrows, from greatest to least, on the basis of the acceleration of the arrows at the top of their flight.

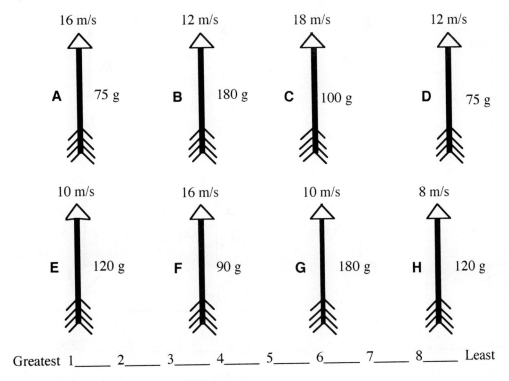

Greatest 1_____ 2_____ 3_____ 4_____ 5_____ 6_____ 7_____ 8_____ Least

All arrows have the same acceleration but not zero. _____

The acceleration at the top is zero for all these. _____

Please carefully explain your reasoning.

How sure were you of your ranking?
Basically Guessed   Sure   Very Sure
   1   2   3   4   5   6   7   8   9   10

[19] T. O'Kuma, D. Maloney
Physics Ranking Tasks

Mechanics

Rocks Thrown Upward—Net Force[20]

Shown below are eight rocks that have been thrown straight up into the air. The rocks all have the same shape, but they have different masses. The rocks are all thrown straight up, but at different speeds. The masses of the rocks and their speeds when released are given in the figures. (We assume for this situation that the effect of air resistance can be ignored.) All start from the same height.

Rank these rocks from greatest to least on the basis of the net force on the rocks after being thrown.

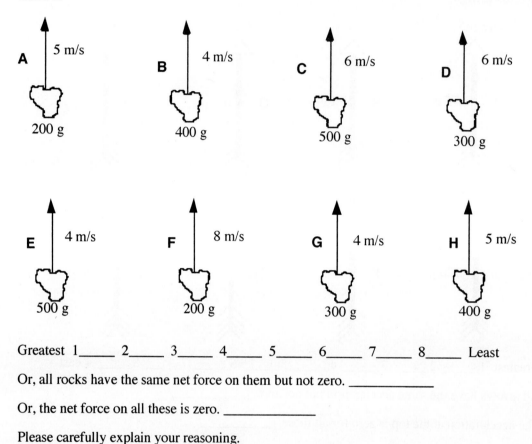

Greatest 1_____ 2_____ 3_____ 4_____ 5_____ 6_____ 7_____ 8_____ Least

Or, all rocks have the same net force on them but not zero. _____

Or, the net force on all these is zero. _____

Please carefully explain your reasoning.

How sure were you of your ranking? (circle one)

Basically Guessed   Sure   Very Sure

1   2   3   4   5   6   7   8   9   10

[20] T. O'Kuma, D. Maloney

Model Rockets Moving Upward—Net Force [21]

The eight figures below depict eight model rockets that have just had their engines turned off. All of the rockets are aimed straight up, but their speeds differ. All of the rockets are the same size and shape, but they carry different loads, so their masses differ. The specific mass and speed for each rocket is given in each figure. (In this situation, we are going to ignore any effect air resistance may have on the rockets.) At the instant when the engines are turned off, the rockets are all at the same height.

Rank these model rockets, from greatest to least, on the basis of the net force on them after the engines have turned off.

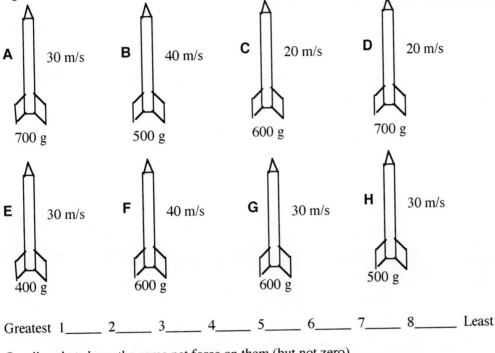

A 30 m/s 700 g
B 40 m/s 500 g
C 20 m/s 600 g
D 20 m/s 700 g
E 30 m/s 400 g
F 40 m/s 600 g
G 30 m/s 600 g
H 30 m/s 500 g

Greatest 1_____ 2_____ 3_____ 4_____ 5_____ 6_____ 7_____ 8_____ Least

Or, all rockets have the same net force on them (but not zero). _____

Or, the net force on all of these is zero. _____

Please carefully explain your reasoning.

How sure were you of your ranking? (circle one)

Basically Guessed					Sure				Very Sure
1	2	3	4	5	6	7	8	9	10

The eight figures below show various situations where blocks of different weights are attached by ropes to rigidly fixed objects or to other blocks, which are attached to fixed objects. The situations differ in a number of ways, as the figures show. The weights of the blocks are given in the figures, as well as the magnitudes and directions of any other forces that may be acting. Our interest is solely in the rope that is designated R in each figure.

Rank these arrangements, from greatest to least, on the basis of the tension in the rope R. That is, put first the arrangement where rope R is under the greatest tension and put last the arrangement where rope R is under the least tension.

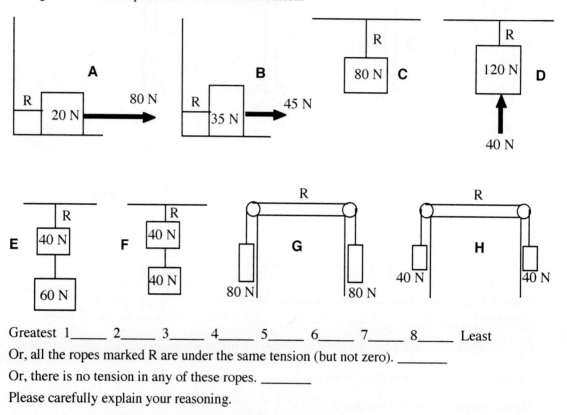

Greatest 1_____ 2_____ 3_____ 4_____ 5_____ 6_____ 7_____ 8_____ Least

Or, all the ropes marked R are under the same tension (but not zero). _____

Or, there is no tension in any of these ropes. _____

Please carefully explain your reasoning.

How sure were you of your ranking? (circle one)

Basically Guessed				Sure				Very Sure	
1	2	3	4	5	6	7	8	9	10

Ball Motion Diagram—Net Force [23]

The following drawings indicate the motion of a ball subject to one or more forces on various surfaces from left to right. Each circle represents the position of the ball at succeeding instants of time. Each time-interval between successive positions is equal, and each ball has the same mass.

Rank the net force on the ball in each case from the highest to the lowest net force based on the figures. Assume the acceleration for each situation to be constant. Note: Positive is to the right and zero is greater than negative.

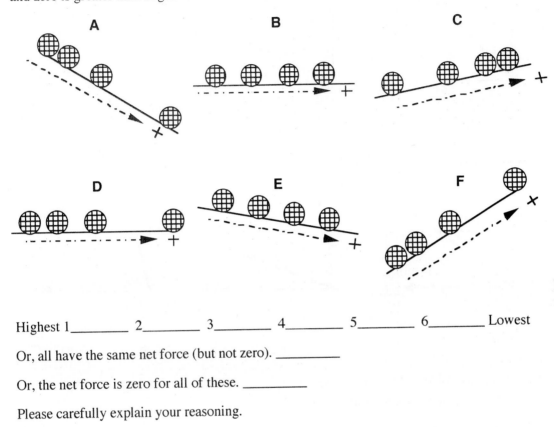

Highest 1_____ 2_____ 3_____ 4_____ 5_____ 6_____ Lowest

Or, all have the same net force (but not zero). _____

Or, the net force is zero for all of these. _____

Please carefully explain your reasoning.

How sure were you of your ranking? (circle one)

Basically Guessed   Sure   Very Sure

1   2   3   4   5   6   7   8   9   10

[23] D. Maloney, T. O'Kuma
Physics Ranking Tasks

Mechanics

The following graphs plot force vs. acceleration for several objects. Rank each situation according to mass. That is, order the situations from the largest to the smallest mass that the force is acting upon. All graphs have the same scale for each respective axis.

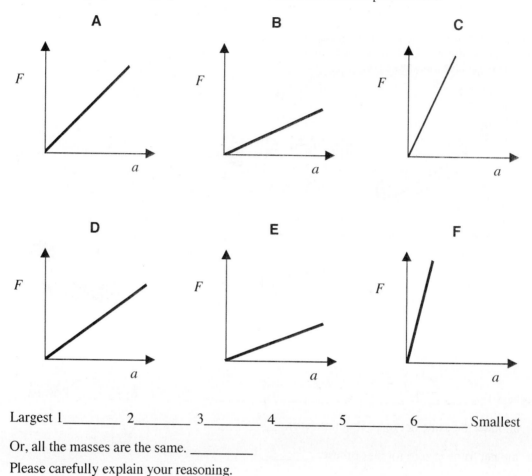

Largest 1_____ 2_____ 3_____ 4_____ 5_____ 6_____ Smallest

Or, all the masses are the same. _____

Please carefully explain your reasoning.

How sure were you of your ranking? (circle one)

Basically Guessed   Sure   Very Sure

1   2   3   4   5   6   7   8   9   10

[24] D. Schramme, C. Fang, B. Speers

Each figure below shows two blocks hanging from the ends of a strong but massless string that passes over a frictionless pulley. In each figure, the block on the left is more massive than the block on the right, so the block on the left accelerates down, and the block on the right accelerates up. The mass of each block is given in the figures.

Rank the figures from greatest to least on the basis of the tension in the string for the system of blocks.

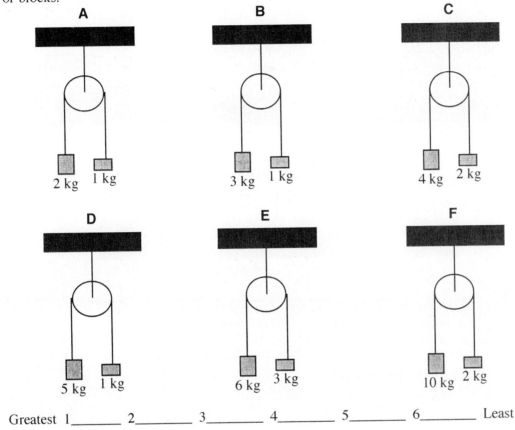

Greatest 1_____ 2_____ 3_____ 4_____ 5_____ 6_____ Least

Or, all of the tensions will be the same. _____

Please carefully explain your reasoning.

How sure were you of your ranking? (circle one)
Basically Guessed   Sure   Very Sure
1   2   3   4   5   6   7   8   9   10

Ropes Pulling Boxes—Acceleration [26]

The figures below show boxes that are being pulled by ropes along frictionless surfaces, accelerating toward the left. All of the boxes are identical. The pulling force applied to the left-most rope is the same in each figure. As you can see, some of the boxes are pulled by ropes attached to the box in front of them.

Rank the masses from greatest to least on the basis of the acceleration of the masses.

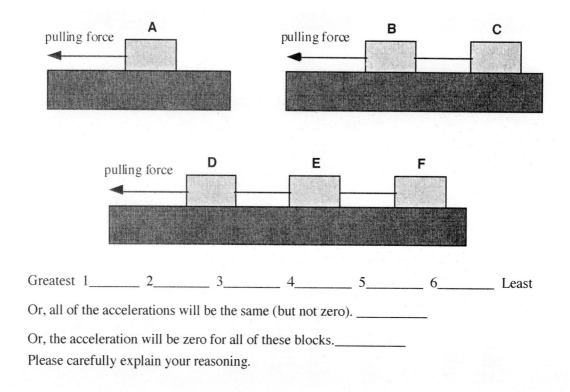

Greatest 1_____ 2_____ 3_____ 4_____ 5_____ 6_____ Least

Or, all of the accelerations will be the same (but not zero). _____

Or, the acceleration will be zero for all of these blocks._____

Please carefully explain your reasoning.

How sure were you of your ranking? (circle one)

Basically Guessed   Sure   Very Sure

1   2   3   4   5   6   7   8   9   10

---

[26] S. Loucks

The figures below show boxes that are being pulled by ropes along frictionless surfaces, accelerating toward the left. All of the boxes are identical, and the acceleration is the same in each figure. As you can see, some of the boxes are pulled by ropes attached to the box in front of them.

Rank the ropes from greatest to least on the basis of the tension in the rope.

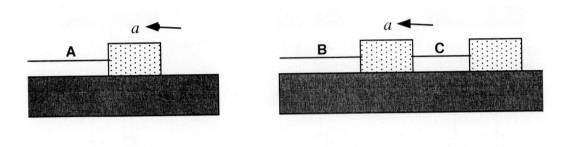

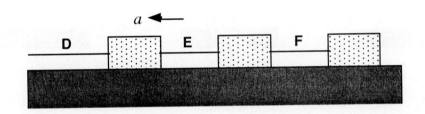

Greatest 1_____ 2_____ 3_____ 4_____ 5_____ 6_____ Least

Or, all of the tensions will be the same. _____

Please carefully explain your reasoning.

How sure were you of your ranking? (circle one)
Basically Guessed   Sure   Very Sure
1   2   3   4   5   6   7   8   9   10

Each figure below shows two blocks hanging from the ends of a strong but massless string, which passes over a frictionless pulley. In each figure, the block on the left is more massive than the block on the right, so the block on the left accelerates down, and the block on the right accelerates up. The mass of each block is given in the figures.

Rank the figures from greatest to least on the basis of the net force that accelerates the system of blocks.

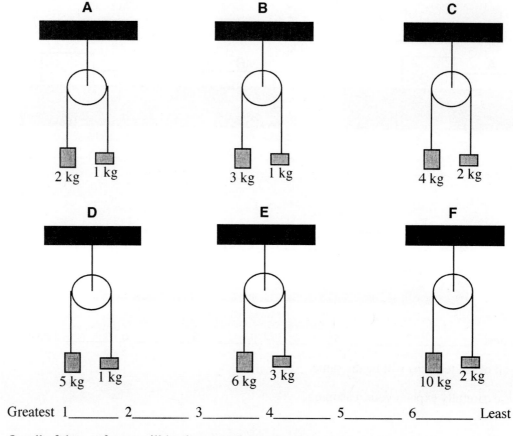

Greatest 1_____ 2_____ 3_____ 4_____ 5_____ 6_____ Least

Or, all of the net forces will be the same (but not zero). ___

Or, the net force is zero for all of these. ____

Please carefully explain your reasoning.

How sure were you of your ranking? (circle one)

Basically Guessed   Sure   Very Sure

1   2   3   4   5   6   7   8   9   10

Moving Car and Boat Trailer—Force Difference [29]

Rank, from greatest to least, on the basis of the difference between the strength (magnitude) of the force the car exerts on the boat trailer, and the strength of the force the boat trailer exerts on the car. All the boat trailers and cars are identical, but the boat trailers have different loads, so the boat trailers masses vary.

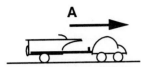

A

$m = 1000$ kg $v_f = 20$ m/s

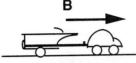

B

$m = 2000$ kg $v_f = 20$ m/s

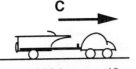

C

$m = 1000$ kg $v_f = 40$ m/s

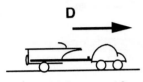

D

$m = 4000$ kg $v_f = 10$ m/s

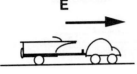

E

$m = 2000$ kg $v_f = 10$ m/s

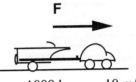

F

$m = 1000$ kg $v_f = 10$ m/s

Greatest 1_____ 2_____ 3_____ 4_____ 5_____ 6_____ Least

Or, the differences between the two forces are the same in each situation. _____

Please carefully explain your reasoning.

How sure were you of your ranking? (circle one)

Basically Guessed				Sure				Very Sure	
1	2	3	4	5	6	7	8	9	10

[29] P. Golden, A. Dickison, D. Maloney, T. O'Kuma, C. Hieggelke

Rank from greatest to least on the basis of the difference between the strength (magnitude) of the force the car exerts on the boat trailer and the strength of the force the trailer exerts on the car during the period when the boat trailers are accelerating from rest to the given final speeds. All the trailers and cabs are identical, but the boat trailers have different loads, so the boat trailer masses vary.

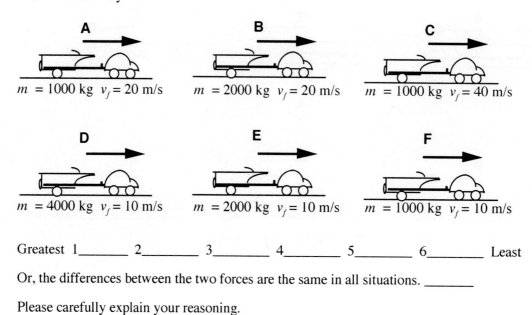

A
$m = 1000$ kg $v_f = 20$ m/s

B
$m = 2000$ kg $v_f = 20$ m/s

C
$m = 1000$ kg $v_f = 40$ m/s

D
$m = 4000$ kg $v_f = 10$ m/s

E
$m = 2000$ kg $v_f = 10$ m/s

F
$m = 1000$ kg $v_f = 10$ m/s

Greatest 1_____ 2_____ 3_____ 4_____ 5_____ 6_____ Least

Or, the differences between the two forces are the same in all situations. _____

Please carefully explain your reasoning.

How sure were you of your ranking? (circle one)

Basically Guessed				Sure				Very Sure	
1	2	3	4	5	6	7	8	9	10

[30] P. Golden, A. Dickison, D. Maloney, T. O'Kuma, C. Hieggelke

Car and Boat Trailer on an Incline—Force Difference [31]

Rank from greatest to least on the basis of the difference between the strength (magnitude) of the force the car exerts on the boat trailer and the strength of the force the boat trailer exerts on the car. All the cars are identical, but the boat trailers have different loads, so the boat trailer masses vary as specified on the diagram. All inclines are the same.

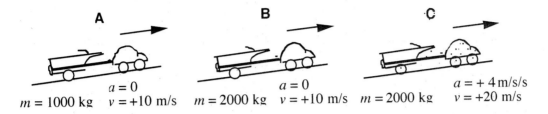

A
$a = 0$
$m = 1000$ kg $v = +10$ m/s

B
$a = 0$
$m = 2000$ kg $v = +10$ m/s

C
$a = +4$ m/s/s
$m = 2000$ kg $v = +20$ m/s

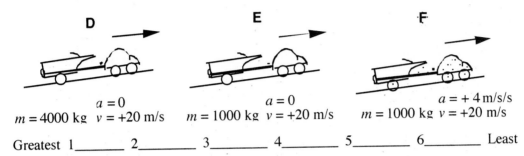

D
$a = 0$
$m = 4000$ kg $v = +20$ m/s

E
$a = 0$
$m = 1000$ kg $v = +20$ m/s

F
$a = +4$ m/s/s
$m = 1000$ kg $v = +20$ m/s

Greatest 1_____ 2_____ 3_____ 4_____ 5_____ 6_____ Least

Or, the differences between the two forces are the same in each situation. _____

Please carefully explain your reasoning.

How sure were you of your ranking? (circle one)

Basically Guessed   Sure   Very Sure
1   2   3   4   5   6   7   8   9   10

[31] P. Golden, A. Dickison, D. Maloney, T. O'Kuma, C. Hieggelke

Forces on Objects on Smooth Surfaces—Velocity Change [32]

Two forces act on an object that is on a frictionless surface, as shown below. Rank these situations from greatest change in velocity to least change in velocity. (Note: All vectors directed to the right are positive, and those to the left are negative. Also, 0 m/s > -10 m/s.)

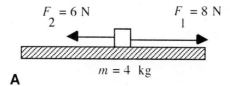

A

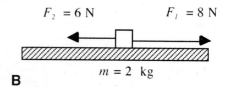

B

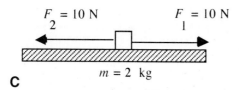

C

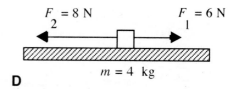

D

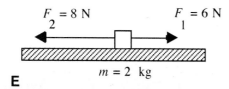

E

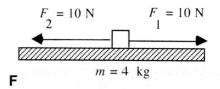

F

Greatest 1_______ 2_______ 3_______ 4_______ 5_______ 6_______ Least

Or, the change in velocity is the same in all cases. _______

Or, the velocity will not change in any of these situations. _______

Please carefully explain your reasoning.

How sure were you of your ranking? (circle one)

Basically Guessed Sure Very Sure

1 2 3 4 5 6 7 8 9 10

[32] R. Krupp

Forces on Objects on Smooth Surfaces—Speed Change [33]

Two forces act on an object that is on a frictionless surface, as shown below. Rank these situations from greatest change in speed to least change in speed.

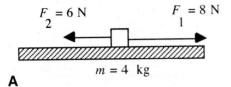

A

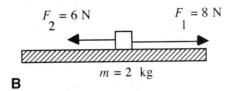

B

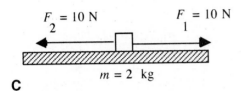

C

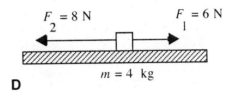

D

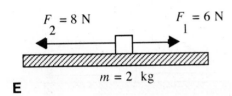

E

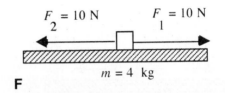

F

Greatest 1_____ 2_____ 3_____ 4_____ 5_____ 6_____ Least

Or, the change in speed is the same in all cases. _____

Or, the speed does not change for any of these cases. _____

Please carefully explain your reasoning.

How sure were you of your ranking? (circle one)

Basically Guessed				Sure					Very Sure
1	2	3	4	5	6	7	8	9	10

Forces on Objects on Rough Surfaces—Velocity Change [34]

Two forces act on identical objects that are on rough surfaces, as shown below. The forces of maximum static and kinetic friction for each case are both 1 N. Rank these situations from greatest change in velocity to least change in velocity. (Note: All vectors directed to the right are positive, and those to the left are negative. Also, 0 m/s > -10 m/s.) All objects start at rest.

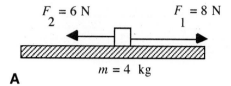

A

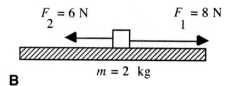

B

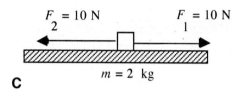

C

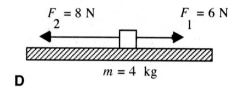

D

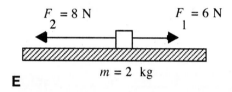

E

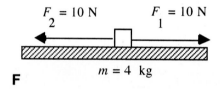

F

Greatest 1_____ 2_____ 3_____ 4_____ 5_____ 6_____ Least

Or, the change in velocity is the same in all cases. _____

Please carefully explain your reasoning.

How sure were you of your ranking? (circle one)

Basically Guessed   Sure   Very Sure
1   2   3   4   5   6   7   8   9   10

Two forces act on identical objects that are on rough surfaces, as shown below. The forces of maximum static and kinetic friction for each case are both 2 N. Rank these situations from greatest change in speed to least change in speed. All objects start at rest.

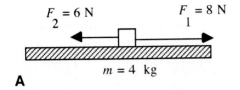

A

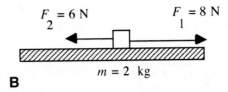

B

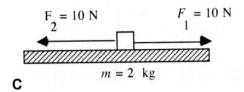

C

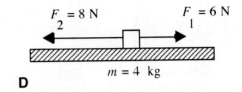

D

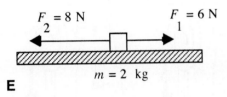

E

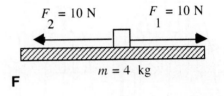

F

Greatest 1_____ 2_____ 3_____ 4_____ 5_____ 6_____ Least

Or, the change in speed is the same in all cases. _____

Please carefully explain your reasoning.

How sure were you of your ranking? (circle one)

Basically Guessed   Sure   Very Sure

1   2   3   4   5   6   7   8   9   10

Person in an Elevator Moving Upward—Scale Weight [36]

The figures below depict situations where a person is standing on a scale in eight identical elevators. Each person weighs 600 N when the elevators are stationary. Each elevator now moves (accelerates) according to the specified arrow that is drawn next to it. In all cases where the elevator is moving, it is moving <u>upward</u>.

Rank the figures, from greatest to least, on the basis of the *scale weight* of each person as registered on each scale. (Use $g = 9.8$ m/s².)

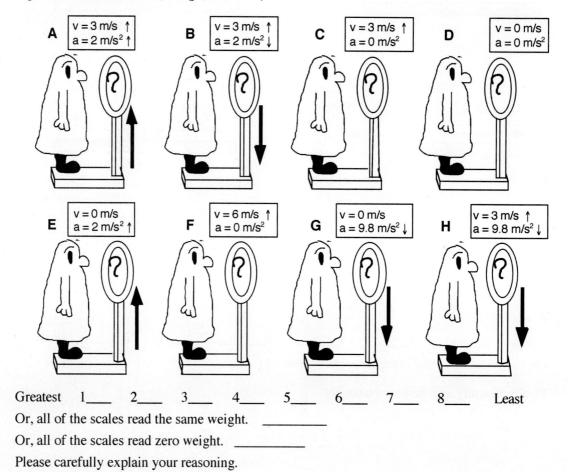

A $v = 3$ m/s ↑ $a = 2$ m/s² ↑

B $v = 3$ m/s ↑ $a = 2$ m/s² ↓

C $v = 3$ m/s ↑ $a = 0$ m/s²

D $v = 0$ m/s $a = 0$ m/s²

E $v = 0$ m/s $a = 2$ m/s² ↑

F $v = 6$ m/s ↑ $a = 0$ m/s²

G $v = 0$ m/s $a = 9.8$ m/s² ↓

H $v = 3$ m/s ↑ $a = 9.8$ m/s² ↓

Greatest   1___   2___   3___   4___   5___   6___   7___   8___   Least

Or, all of the scales read the same weight.   _____

Or, all of the scales read zero weight.   _____

Please carefully explain your reasoning.

How sure were you of your ranking? (circle one)

Basically Guessed   Sure   Very Sure
1   2   3   4   5   6   7   8   9   10

The figures below depict situations where a person is standing on a scale in eight identical elevators. Each person weighs 600 N when the elevators are stationary. Each elevator now moves (accelerates) according to the specified arrow that is drawn next to it. In all cases where the elevator is moving, it is moving <u>downward</u>.

Rank the figures, from greatest to least, on the basis of the *scale weight* of each person as registered on each scale. (Use $g = 9.8$ m/s².)

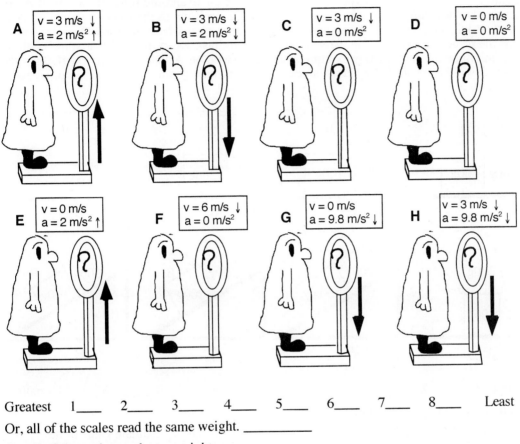

A | $v = 3$ m/s ↓ | $a = 2$ m/s² ↑

B | $v = 3$ m/s ↓ | $a = 2$ m/s² ↓

C | $v = 3$ m/s ↓ | $a = 0$ m/s²

D | $v = 0$ m/s | $a = 0$ m/s²

E | $v = 0$ m/s | $a = 2$ m/s² ↑

F | $v = 6$ m/s ↓ | $a = 0$ m/s²

G | $v = 0$ m/s | $a = 9.8$ m/s² ↓

H | $v = 3$ m/s ↓ | $a = 9.8$ m/s² ↓

Greatest   1___   2___   3___   4___   5___   6___   7___   8___   Least

Or, all of the scales read the same weight. _____

Or, all of the scales read zero weight. _____

Please carefully explain your reasoning.

How sure were you of your ranking? (circle one)
Basically Guessed   Sure   Very Sure
1   2   3   4   5   6   7   8   9   10

Two Blocks at Rest—Force Difference [38]

Shown below are eight arrangements of two wooden blocks. There are two different mass blocks, either 100 g or 200 g. In all of the arrangements, the blocks are in contact, that is, they are touching each other. Also, in all of the arrangements the blocks are at rest, i.e., they are not moving. As you can see, one of the blocks given in each arrangement is labeled **A,** and the other is labeled **B**. The mass of each block is given in the figures.

Rank these arrangements from largest to smallest on the basis of the difference of the strengths (magnitudes) of the forces between the force **A** exerts on **B** and the force **B** exerts on **A**. In other words, the arrangement where the force **A** exerts on **B** minus the force **B** exerts on **A** is the largest will rank first. In the same way, the arrangement where the force **A** exerts on **B** minus the force **B** exerts on **A** is the smallest will rank last. Keep in mind that some of these values might be negative. If **B** is exerting a stronger force on **A** than **A** exerts on **B**, then the difference will be negative. Negative values are smaller than positive values or zero. (A force is a push or a pull.)

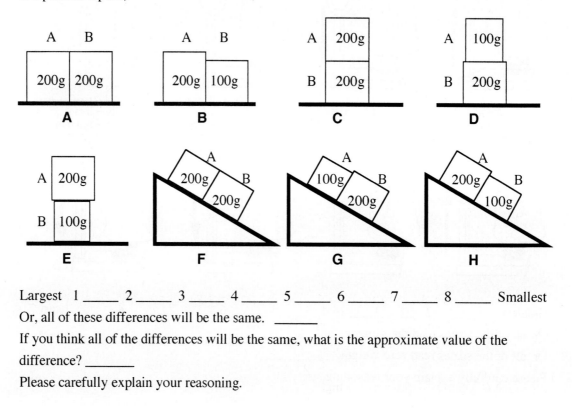

Largest 1 _____ 2 _____ 3 _____ 4 _____ 5 _____ 6 _____ 7 _____ 8 _____ Smallest

Or, all of these differences will be the same. _____

If you think all of the differences will be the same, what is the approximate value of the difference? _____

Please carefully explain your reasoning.

How sure were you of your ranking? (circle one)

Basically Guessed					Sure				Very Sure
1	2	3	4	5	6	7	8	9	10

[38] D. Maloney

Shown below are eight arrangements of two wooden blocks both moving left to right at 2 m/s. There are two different mass blocks, either 100 g or 200 g. In all of the arrangements, the blocks are in contact, that is, they are touching each other. As you can see, one of the blocks given in each arrangement is labeled **A**, and the other is labeled **B**. The mass of each block is given in the figures.

Rank these arrangements from largest to smallest on the basis of the difference of the strengths (magnitudes) of the forces between the force **A** exerts on **B** and the force **B** exerts on **A**. In other words, the arrangement where the force **A** exerts on **B** minus the force **B** exerts on **A** is the largest will rank first. In the same way the arrangement where the force **A** exerts on **B** minus the force **B** exerts on **A** is the smallest will rank last. Keep in mind that some of these values might be negative. If **B** is exerting a stronger force on **A** than **A** exerts on **B**, then the difference will be negative. Negative values are smaller than positive values or zero. (A force is a push or a pull.)

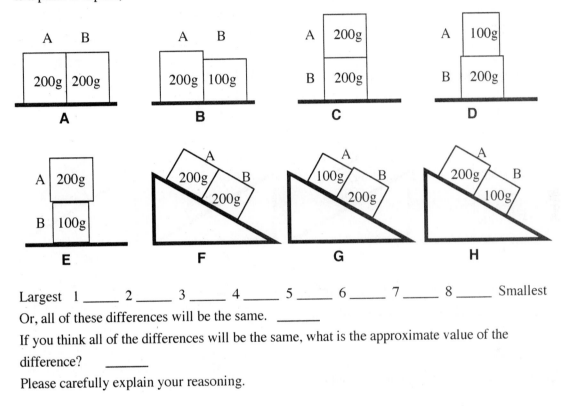

Largest 1 _____ 2 _____ 3 _____ 4 _____ 5 _____ 6 _____ 7 _____ 8 _____ Smallest

Or, all of these differences will be the same. _____

If you think all of the differences will be the same, what is the approximate value of the difference? _____

Please carefully explain your reasoning.

How sure were you of your ranking? (circle one)
Basically Guessed   Sure   Very Sure
   1   2   3   4   5   6   7   8   9   10

Shown below are eight arrangements of two wooden blocks both moving left to right at 2 m/s and accelerating in the same direction at 3 m/s². There are two different mass blocks, either 100 g or 200 g. In all of the arrangements, the blocks are in contact, that is, they are touching each other. As you can see, one of the blocks given in each arrangement is labeled **A**, and the other is labeled **B**. The mass of each block is given in the figures.

Rank these arrangements from largest to smallest on the basis of the difference of the strengths (magnitudes) of the forces between the force **A** exerts on **B** and the force **B** exerts on **A**. In other words, the arrangement where the force **A** exerts on **B** minus the force **B** exerts on **A** is the largest will rank first. In the same way the arrangement where the force **A** exerts on **B** minus the force **B** exerts on **A** is the smallest will rank last. Keep in mind that some of these values might be negative. If **B** is exerting a stronger force on **A** than **A** exerts on **B**, then the difference will be negative. Negative values are smaller than positive values or zero. (A force is a push or a pull.)

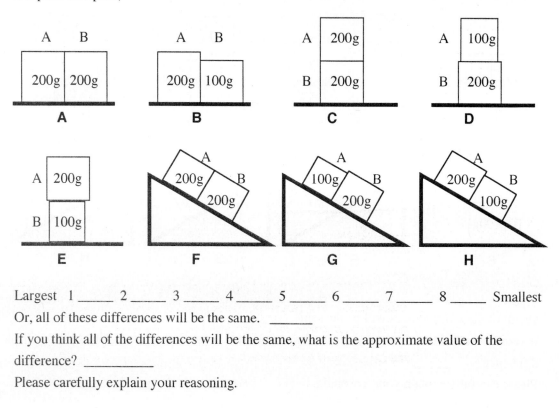

Largest 1 _____ 2 _____ 3 _____ 4 _____ 5 _____ 6 _____ 7 _____ 8 _____ Smallest

Or, all of these differences will be the same. _____

If you think all of the differences will be the same, what is the approximate value of the difference? _____

Please carefully explain your reasoning.

How sure were you of your ranking? (circle one)

Basically Guessed   Sure   Very Sure

1   2   3   4   5   6   7   8   9   10

[40] D. Maloney, C. Hieggelke

Horizontal Arrows at Different Distances—Force [41]

The figures below show arrows that have been shot from bows. All of the arrows are identical, and they are moving horizontally to the right. The arrows are at different points in their paths from the bows to the targets. The distances the arrows have traveled in reaching the points shown are given in the figures. Also given in the figures are the speeds the arrows have at the points shown.

Rank these situations, from greatest to least, on the basis of the rightward pointing force, i.e., the force acting in the direction the arrow is moving, acting on each arrow at the point shown. That is, put first the arrow with the largest horizontal force acting on it, and put last the arrow with the smallest horizontal force. (A force is a push or pull.) We are ignoring any effects of air in these situations.

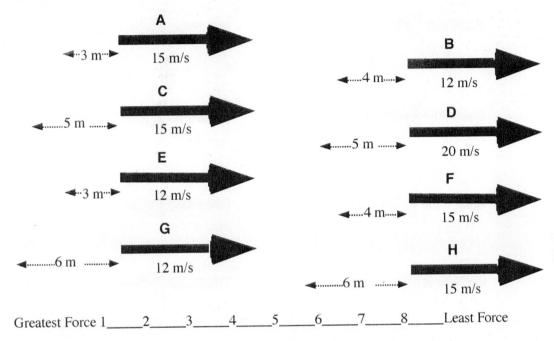

Greatest Force 1_____2_____3_____4_____5_____6_____7_____8_____Least Force

Or, all of these arrows have the same horizontal force acting upon them. _____

If you check this answer, what is your estimate of the strength of the force? _____ Please

carefully explain your reasoning.

How sure were you of the reasoning you used? (circle one)
Basically Guessed   Sure   Very Sure
1   2   3   4   5   6   7   8   9   10

Horizontal Arrows at Different Times—Force [42]

The figures below show arrows that have been shot from bows. All of the arrows are identical, and they are moving horizontally to the right. The arrows are at different points in their paths from the bows to the targets. The times since being shot vary for the arrows. These times are given in the figures. Also given in the figures are the speeds the arrows have at the specified times.

Rank these situations, from greatest to least, on the basis of the rightward pointing force; i.e., the force acting in the direction the arrow is moving, acting on each arrow at the point shown. That is, put first the arrow with the largest horizontal force acting on it, and put last the arrow with the smallest horizontal force. (A force is a push or pull.) We are ignoring any effects of air in these situations.

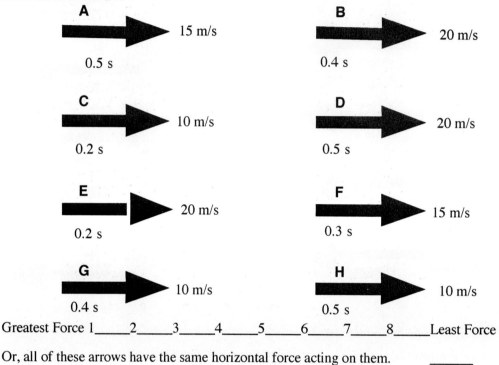

Greatest Force 1_____2_____3_____4_____5_____6_____7_____8_____Least Force

Or, all of these arrows have the same horizontal force acting on them. _____

If you check this answer what is your estimate of the strength of the force? _____

Please carefully explain your reasoning.

How sure were you of the reasoning you used? (circle one)

Basically Guessed   Sure   Very Sure
1   2   3   4   5   6   7   8   9   10

The figures below show arrows that have been shot from bows. All of the arrows are identical, and they are moving horizontally to the right. The arrows are at different points in their paths from the bows to the targets. The distances the arrows have traveled and the times required are given in the figures.

Rank these situations, from greatest to least, on the basis of the rightward pointing force; i.e., the force acting in the direction the arrow is moving, acting on each arrow at the point shown. That is, put first the arrow with the largest horizontal force acting on it, and put last the arrow with the smallest horizontal force. (A force is a push or pull.) We are ignoring any effects of air in these situations.

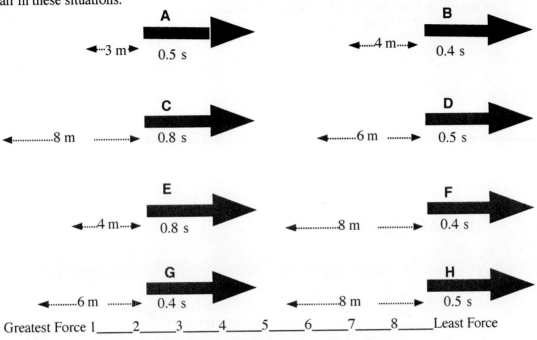

Greatest Force 1_____2_____3_____4_____5_____6_____7_____8_____Least Force

Or, all of these arrows have the same horizontal force acting on them._____

If you check this answer, what is your estimate of the strength of the force?_____

Please carefully explain your reasoning.

How sure were you of the reasoning you used? (circle one)

Basically Guessed					Sure				Very Sure
1	2	3	4	5	6	7	8	9	10

Projectile and Other Two-Dimensional Motion Ranking Tasks

Water Over a Waterfall—Time to Reach Ground [44]

Pictured below are six waterfalls all of which have the same amount of water flowing over them. The waterfalls differ in height and in the speed of the water as it goes over the edge. The specific values of the heights and speeds are given in the figures.

Rank these situations from longest to shortest based on how long it takes the water to go from the top of the falls to the bottom. That is, put first the situation where it takes the water the most time to go from the top of the falls to the bottom, and put last the one that takes the least time.

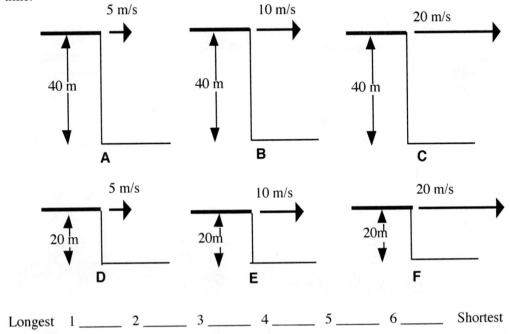

Longest 1 _____ 2 _____ 3 _____ 4 _____ 5 _____ 6 _____ Shortest

Or, water from all of the waterfalls reaches the bottom at the same time. _____

Please carefully explain your reasoning.

How sure were you of your ranking? (circle one)

Basically Guessed					Sure				Very Sure
1	2	3	4	5	6	7	8	9	10

Pictured below are eight arrows that have been shot horizontally, i.e., straight out, by archers on platforms. All of the arrows are identical, but they have been shot at different speeds from platforms of varying height. Specific values for the speeds and of varying heights are given in the figures. All of the arrows miss the targets and hit the ground.

Rank these arrows, from longest to shortest, on the basis of how long it takes the arrows to hit the ground. That is, put first the arrow that will take the longest time from being shot to hitting the ground, and put last the arrow that will take the shortest time.

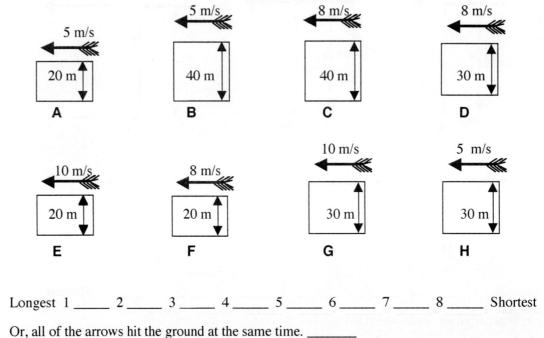

Longest 1 _____ 2 _____ 3 _____ 4 _____ 5 _____ 6 _____ 7 _____ 8 _____ Shortest

Or, all of the arrows hit the ground at the same time. _____

Please carefully explain your reasoning.

How sure were you of your ranking? (circle one)
Basically Guessed   Sure   Very Sure
1   2   3   4   5   6   7   8   9   10

Rifle Shots—Time to Hit Ground [46]

The eight figures below show rifles that are being fired horizontally, i.e., straight out, off platforms. The bullets fired from the rifles are all identical, but the rifles propel the bullets at different speeds. The specific speed of each bullet and the height of each platform is given. All of the bullets miss the targets and hit the ground.

Rank these bullets, from longest to shortest, on the basis of how long it takes a bullet to hit the ground. That is, put first the bullet that will take the longest time from being fired to hitting the ground, and put last the bullet that will take the shortest time.

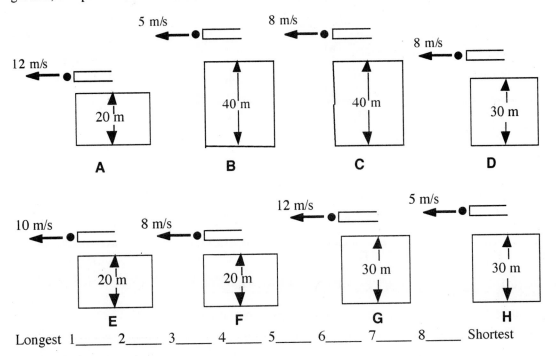

Longest 1_____ 2_____ 3_____ 4_____ 5_____ 6_____ 7_____ 8_____ Shortest

Or, all of the bullets reach the ground at the same. _____

Please carefully explain your reasoning.

How sure were you of your ranking? (circle one)

Basically Guessed					Sure			Very Sure	
1	2	3	4	5	6	7	8	9	10

Toy Trucks Rolling Off Tables—Time in Air [47]

Shown in the eight figures below are toy trucks that are rolling along on tables. All of the trucks are identical. The tables they are on vary in height, and the speeds of the trucks vary. Specific values for the heights and speeds are given in the figures. Each truck rolls off the end of its table.

Rank these trucks from longest to shortest, based on how long the trucks are in the air. That is, put first the truck that takes the longest time to go from the edge of the table to the ground, and put last the truck that takes the least time.

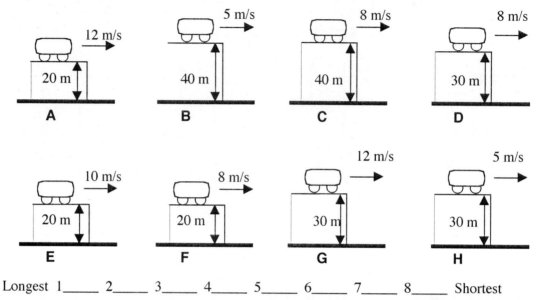

Longest 1_____ 2_____ 3_____ 4_____ 5_____ 6_____ 7_____ 8_____ Shortest

Or, all of the toy trucks are in the air the same amount of time. _____

Please carefully explain your reasoning.

How sure were you of your ranking? (circle one)

Basically Guessed   Sure   Very Sure

1   2   3   4   5   6   7   8   9   10

---

[47] D. Maloney

Spheres Thrown Horizontally Off Cliffs—Time to Hit Ground [48]

The eight figures below show spheres that have been thrown horizontally, i.e., straight out, off cliffs. All of the spheres are the same size and have the same mass, but they are thrown at different speeds off cliffs of different heights. The specific speeds and heights are given in each figure.

Rank these spheres from longest to shortest on the basis of how long they take to reach the ground. That is, put first the sphere that takes the most time in the air, and put last the sphere that takes the shortest time in the air.

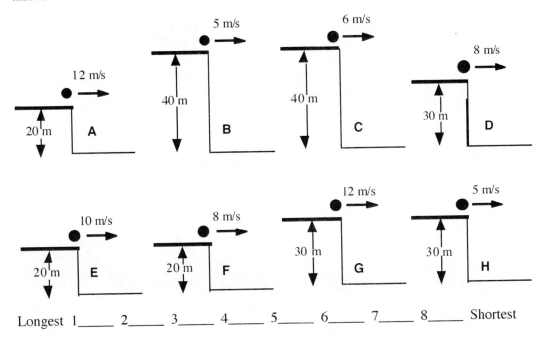

Longest 1_____ 2_____ 3_____ 4_____ 5_____ 6_____ 7_____ 8_____ Shortest

Or, all of the spheres take the same time in the air. _____

Please carefully explain your reasoning.

How sure were you of your ranking? (circle one)

Basically Guessed   Sure   Very Sure
1   2   3   4   5   6   7   8   9   10

The eight figures below show arrows that have been shot into the air. All of the arrows were shot at the same angle and are the same size and shape. The arrows are made of different materials so they have different masses, and they have different speeds as they leave the bows. The values for each arrow are given in the figures. (We assume for this situation that the effect of air resistance can be ignored.) All start from same height.

Rank these arrows from greatest to least on the basis of the maximum heights the arrows reach.

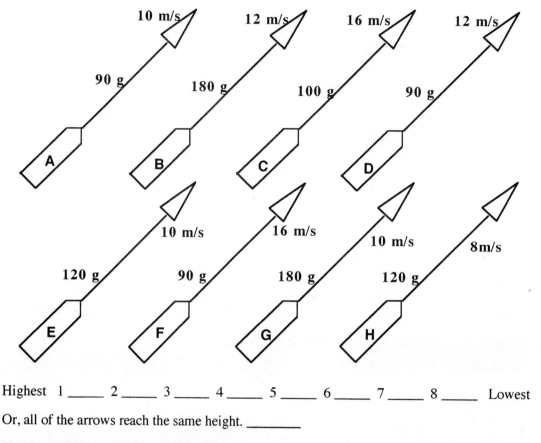

Highest 1 _____ 2 _____ 3 _____ 4 _____ 5 _____ 6 _____ 7 _____ 8 _____ Lowest

Or, all of the arrows reach the same height. _____

Please carefully explain your reasoning.

How sure were you of your ranking? (circle one)
Basically Guessed   Sure   Very Sure
1   2   3   4   5   6   7   8   9   10

Rock Throw—Maximum Heights [50]

Shown below are eight rocks that have been thrown into the air. The rocks all have the same shape, but they have different masses. The rocks are all thrown at the same angle, but at different speeds. The masses of the rocks and their speeds, when released, are given in the figures. (We assume for this situation that the effect of air resistance can be ignored.) All start from the same height.

Rank these rocks from greatest to least on the basis of the maximum heights the rocks reach.

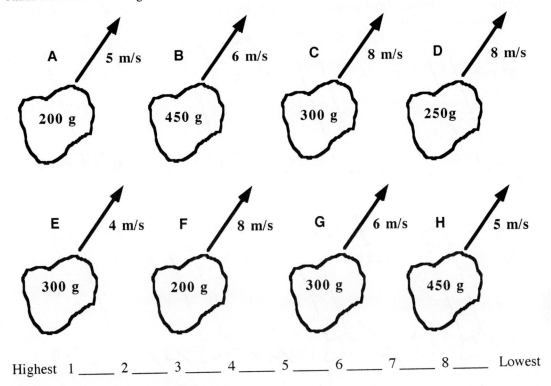

Highest 1 _____ 2 _____ 3 _____ 4 _____ 5 _____ 6 _____ 7 _____ 8 _____ Lowest

Or, all rocks reach the same height. _____

Please carefully explain your reasoning.

How sure were you of your ranking? (circle one)

Basically Guessed					Sure				Very Sure
1	2	3	4	5	6	7	8	9	10

Model Rockets Fired at an Angle—Horizontal Speed at Top [51]

The eight figures below depict eight model rockets that have just had their engines turned off. All of the rockets are aimed upward at the same angle, but their speeds differ. All of the rockets are the same size and shape, but they carry different loads, so their masses differ. The specific mass and speed for each rocket is given in each figure. (In this situation we are going to ignore any effect air resistance may have on the rockets.) At the instant when the engines are turned off, the rockets are all at the same height.

Rank these model rockets from greatest to least on the basis of the horizontal speed at the top (at the maximum height).

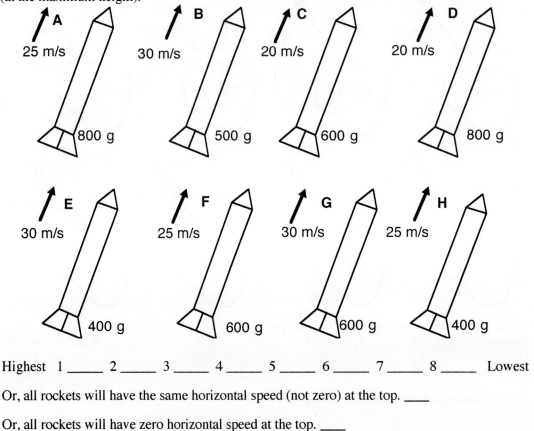

Highest 1 _____ 2 _____ 3 _____ 4 _____ 5 _____ 6 _____ 7 _____ 8 _____ Lowest

Or, all rockets will have the same horizontal speed (not zero) at the top. ____

Or, all rockets will have zero horizontal speed at the top. ____

Please carefully explain your reasoning.

How sure were you of your ranking? (circle one)

Basically Guessed					Sure				Very Sure
1	2	3	4	5	6	7	8	9	10

[51] D. Maloney

Cannon Shots—Acceleration at the Top [52]

The figures below depict eight cannons shooting shells into the air. All of the cannons are aimed at the same angle of 35 degrees. All of the cannons are identical. The shells are all the same size and shape, but the masses of the shells, as well as their speeds as they leave the cannons, are different. The values of these variables are specified in the figures.

Rank these situations from highest to lowest on the basis of the acceleration at the highest point reached by the shells. (We assume for this situation that the effect of air resistance can be ignored.)

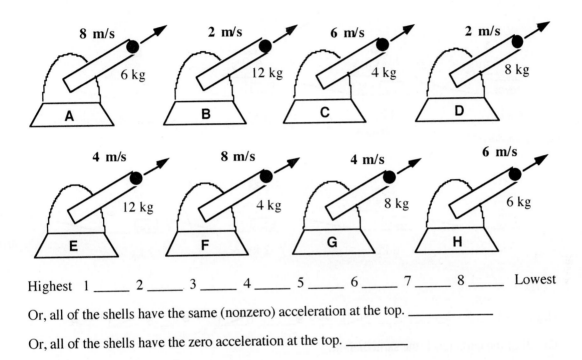

Highest 1 _____ 2 _____ 3 _____ 4 _____ 5 _____ 6 _____ 7 _____ 8 _____ Lowest

Or, all of the shells have the same (nonzero) acceleration at the top. _____

Or, all of the shells have the zero acceleration at the top. _____

Please carefully explain your reasoning.

How sure were you of your ranking? (circle one)

Basically Guessed					Sure				Very Sure
1	2	3	4	5	6	7	8	9	10

Carts on Incline—Height After Leaving Incline [53]

Shown below are diagrams of situations where carts, initially moving along a horizontal surface, run up onto short inclines. The carts go up off the inclines. All of the inclines have the same angle, and they are very smooth, so there is essentially no friction. The masses of the carts and their speeds when they reach the inclines are given in the figures.

Rank the situations in order of how high the carts go after leaving the incline. That is, order the situations from the one where the cart goes highest (greatest vertical distance) to that where the cart goes lowest.

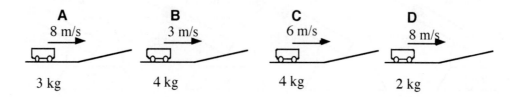

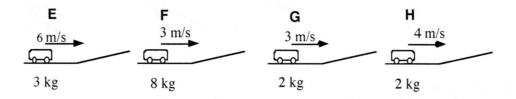

Highest 1_____ 2_____ 3_____ 4_____ 5_____ 6_____ 7_____ 8_____ Lowest

Or, all of the carts reach the same height. _________

Please carefully explain your reasoning.

How sure were you of your ranking? (circle one)

Basically Guessed					Sure				Very Sure
1	2	3	4	5	6	7	8	9	10

[53] D. Maloney

Projectile—Horizontal Distance [54]

The pictures below depict cannonballs of two different masses projected upward and forward. The cannonballs are projected at various angles above the horizontal, but all are projected with the same horizontal component of velocity.

Rank according to the horizontal distance traveled by the balls.

A
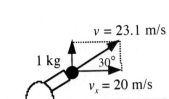

B
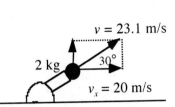

C
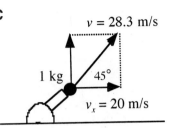

D
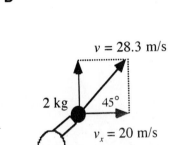

E
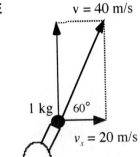

F
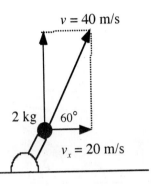

Largest 1_____ 2_____ 3_____ 4_____ 5_____ 6_____ Smallest

All distances traveled are the same. _____

Please carefully explain your reasoning.

How sure were you of your ranking? (circle one)

Basically Guessed   Sure   Very Sure

1   2   3   4   5   6   7   8   9   10

Projectile—Time in Air [55]

The pictures below depict cannonballs of two different masses projected upward and forward. The cannonballs are projected at various angles above the horizontal, but all are projected with the same vertical component of velocity.

Rank according to the time the balls are in the air.

A

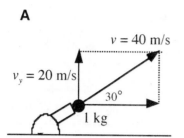

B

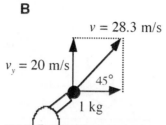

C

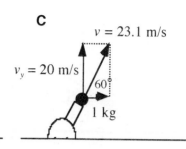

D

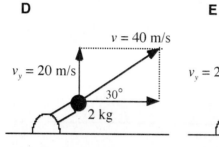

E

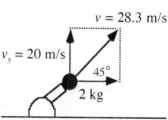

F

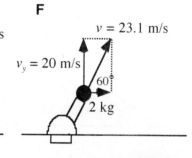

Largest 1_____ 2_____ 3_____ 4_____ 5_____ 6_____ Smallest

All times are the same. _____

Please carefully explain your reasoning.

How sure were you of your ranking? (circle one)
Basically Guessed   Sure   Very Sure
1   2   3   4   5   6   7   8   9   10

Work—Energy Ranking Tasks

In a western movie, a confederate raiding party stopped a runaway boxcar carrying gold by using many ropes tied to trees. Given below are six boxcars that are moving along horizontal railroads at specified speeds. Also given are the masses of the boxcars. All of the boxcars are the same size and shape, but they are carrying loads with different masses. All of these boxcars are going to be stopped by plowing through a large number of these secured ropes. All of the boxcars need to be stopped in the same distance.

Rank these situations from greatest to least on the basis of the strength of the forces that will be needed to stop the boxcars in the same distance. That is, put first the boxcar on which the strongest force will have to be applied to stop it in x meters, and put last the boxcar on which the weakest force will be applied to stop the boxcar in the same distance.

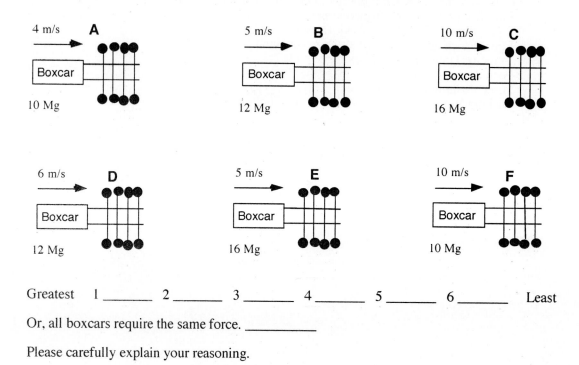

Greatest 1 _____ 2 _____ 3 _____ 4 _____ 5 _____ 6 _____ Least

Or, all boxcars require the same force. _____

Please carefully explain your reasoning.

How sure were you of your ranking? (circle one)
Basically Guessed   Sure   Very Sure
1   2   3   4   5   6   7   8   9   10

Cars and Barriers—Stopping Force in Same Distance I [57]

Given below are eight cars that are moving along horizontal roads at specified speeds. Also given are the masses of the cars. All of the cars are the same size and shape, but they are carrying loads with different masses. All of these cars are going to be stopped by plowing into barrel barriers. All of the cars are going to be stopped in the same distance.

Rank these situations from greatest to least on the basis of the strength of the forces that will be needed to stop the cars in the same distance. That is, put first the car on which the strongest force will have to be applied to stop it in x meters, and put last the car on which the weakest force will be applied to stop the car in the same distance.

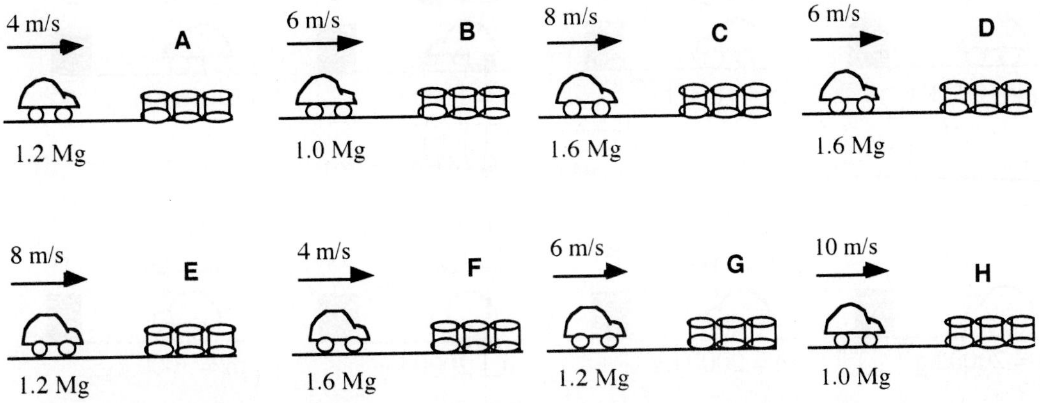

Greatest 1_____ 2_____ 3_____ 4_____ 5_____ 6_____ 7_____ 8_____ Least

Or, all cars require the same force. ________

Please carefully explain your reasoning.

How sure were you of your ranking? (circle one)

Basically Guessed					Sure			Very Sure	
1	2	3	4	5	6	7	8	9	10

[57] T. O'Kuma

Cars and Barriers—Stopping Distance with the Same Force[58]

Shown below are eight cars that are moving along horizontal roads at specified speeds. Also given are the masses of the cars. All of the cars are the same size and shape, but they are carrying loads with different masses. All of these cars are going to be stopped by plowing into identical barriers. All of the cars are going to be stopped by the same constant force by the barrier.

Rank these situations from greatest to least on the basis of the stopping distance that will be needed to stop the cars with the same force. That is, put first the car that requires the longest stopping distance and put last the car that requires the shortest distance to stop the car with the same force.

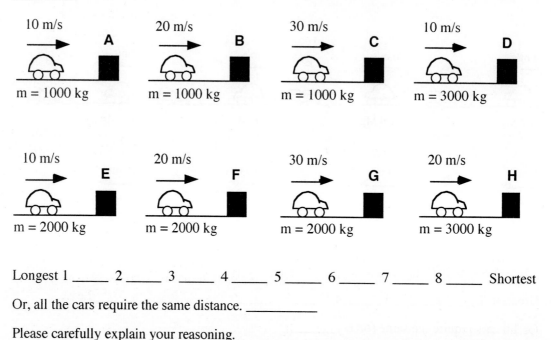

Longest 1 _____ 2 _____ 3 _____ 4 _____ 5 _____ 6 _____ 7 _____ 8 _____ Shortest

Or, all the cars require the same distance. _____

Please carefully explain your reasoning.

How sure were you of your ranking? (circle one)

Basically Guessed				Sure				Very Sure	
1	2	3	4	5	6	7	8	9	10

[58] T. O'Kuma, D. Maloney, C. Hieggelke

The eight situations below show *before* and *after* "snapshots" of a car's velocity. Rank these situations, in terms of work done on the car, from most positive to most negative, to create these changes in velocity for the same distance traveled. All cars have the same mass. Negative numbers, if any, rank lower than positive ones (-20 m/s < -10 m/s < 0< 5).

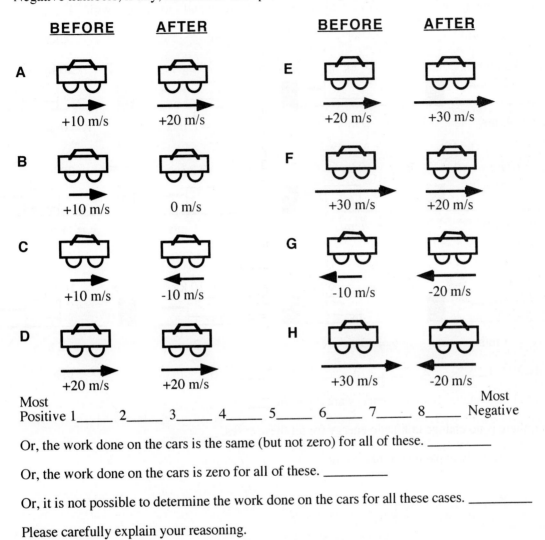

Most
Positive 1_____ 2_____ 3_____ 4_____ 5_____ 6_____ 7_____ 8_____ Most
Negative

Or, the work done on the cars is the same (but not zero) for all of these. _____

Or, the work done on the cars is zero for all of these. _____

Or, it is not possible to determine the work done on the cars for all these cases. _____

Please carefully explain your reasoning.

How sure were you of your ranking? (circle one)

Basically Guessed				Sure				Very Sure	
1	2	3	4	5	6	7	8	9	10

[59] J. Cole, D. Maloney
Physics Ranking Tasks

Mechanics

Bouncing Cart—Change in Kinetic Energy [60]

A cart with a spring plunger runs into a fixed barrier. The mass of the cart, its velocity just before impact with the barrier, and its velocity right after collision are given in each figure.

Rank the change in kinetic energy for each cart from the greatest change in kinetic energy to the least change in kinetic energy (+ direction is to the right and - to the left, with –4 < -2).

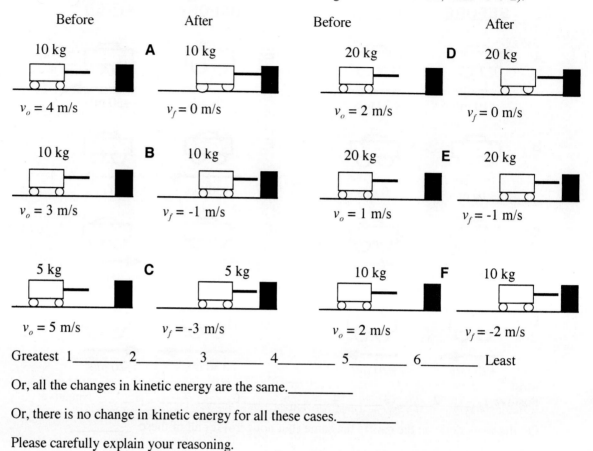

Before	After	Before	After
10 kg	**A** 10 kg	20 kg	**D** 20 kg
$v_o = 4$ m/s	$v_f = 0$ m/s	$v_o = 2$ m/s	$v_f = 0$ m/s
10 kg	**B** 10 kg	20 kg	**E** 20 kg
$v_o = 3$ m/s	$v_f = -1$ m/s	$v_o = 1$ m/s	$v_f = -1$ m/s
5 kg	**C** 5 kg	10 kg	**F** 10 kg
$v_o = 5$ m/s	$v_f = -3$ m/s	$v_o = 2$ m/s	$v_f = -2$ m/s

Greatest 1_____ 2_____ 3_____ 4_____ 5_____ 6_____ Least

Or, all the changes in kinetic energy are the same._____

Or, there is no change in kinetic energy for all these cases._____

Please carefully explain your reasoning.

How sure were you of your ranking? (circle one)
Basically Guessed   Sure   Very Sure
1   2   3   4   5   6   7   8   9   10

[60] T. O'Kuma, C. Hieggelke
Physics Ranking Tasks

Mechanics

Bouncing Cart—Work Done by the Barrier [61]

A cart with a spring plunger runs into a fixed barrier. The mass of the cart, its velocity just before impact with the barrier, and its velocity right after collision are given in each figure.

Rank the work done *by* the barrier on each cart from the greatest work done by the barrier to the least work done by the barrier (+ direction is to the right and - to the left with $-4 < -2$).

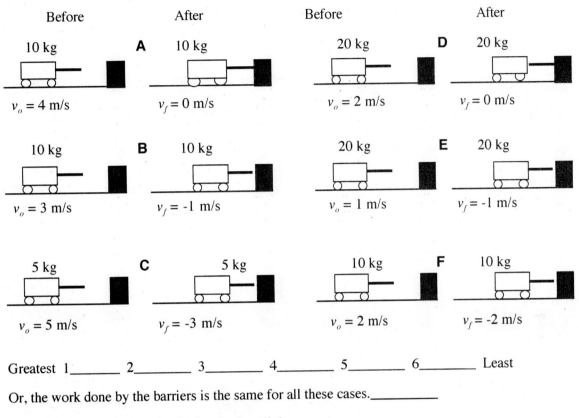

Before		After		Before		After

A
10 kg — 10 kg
$v_o = 4$ m/s $v_f = 0$ m/s

B
10 kg — 10 kg
$v_o = 3$ m/s $v_f = -1$ m/s

C
5 kg — 5 kg
$v_o = 5$ m/s $v_f = -3$ m/s

D
20 kg — 20 kg
$v_o = 2$ m/s $v_f = 0$ m/s

E
20 kg — 20 kg
$v_o = 1$ m/s $v_f = -1$ m/s

F
10 kg — 10 kg
$v_o = 2$ m/s $v_f = -2$ m/s

Greatest 1_____ 2_____ 3_____ 4_____ 5_____ 6_____ Least

Or, the work done by the barriers is the same for all these cases._____

Or, there is no work done by the barrier for all these cases._____

Please carefully explain your reasoning.

Bouncing Cart—Work Done on the Barrier [62]

A cart with a spring plunger runs into a fixed barrier. The mass of the cart, its velocity just before impact with the barrier, and its velocity right after collision are given in each figure.

Rank the work done *on* the barrier by each cart from the greatest work done on the barrier to the least work done on the barrier (+ direction is to the right and - to the left with $-4 < -2$).

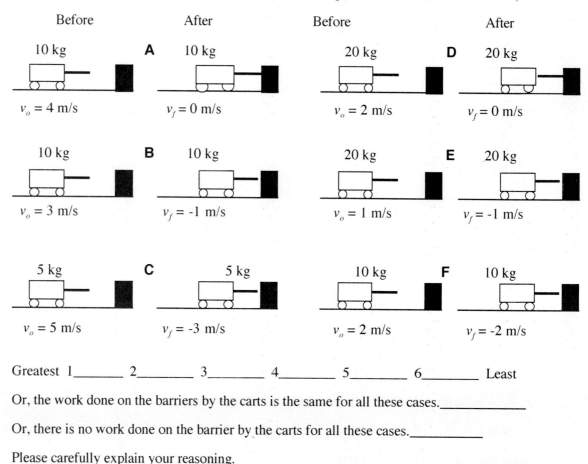

Before	After	Before	After

A 10 kg, $v_o = 4$ m/s → 10 kg, $v_f = 0$ m/s

B 10 kg, $v_o = 3$ m/s → 10 kg, $v_f = -1$ m/s

C 5 kg, $v_o = 5$ m/s → 5 kg, $v_f = -3$ m/s

D 20 kg, $v_o = 2$ m/s → 20 kg, $v_f = 0$ m/s

E 20 kg, $v_o = 1$ m/s → 20 kg, $v_f = -1$ m/s

F 10 kg, $v_o = 2$ m/s → 10 kg, $v_f = -2$ m/s

Greatest 1_____ 2_____ 3_____ 4_____ 5_____ 6_____ Least

Or, the work done on the barriers by the carts is the same for all these cases._____

Or, there is no work done on the barrier by the carts for all these cases._____

Please carefully explain your reasoning.

How sure were you of your ranking? (circle one)

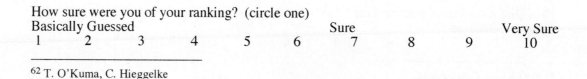

Basically Guessed   Sure   Very Sure

1   2   3   4   5   6   7   8   9   10

Model Rockets—Kinetic Energy [63]

The eight figures below depict eight model rockets that have just had their engines turned off. All of the rockets are aimed straight up, but their speeds differ. All of the rockets are the same size and shape, but they carry different loads, so their masses differ. The mass and speed for each rocket is given in each figure. (In this situation we are going to ignore any effect air resistance may have on the rockets.) At the instant when the engines are turned off, the rockets are all at the same height.

Rank these model rockets from greatest to least on the basis of the kinetic energy they have at the top of their flights.

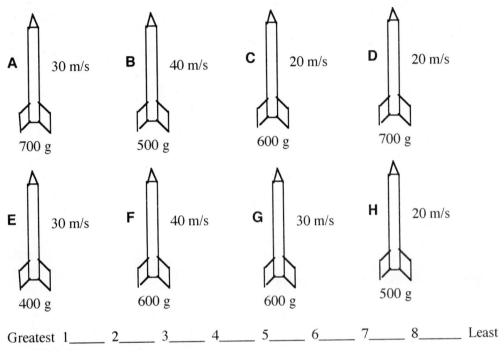

A 30 m/s 700 g

B 40 m/s 500 g

C 20 m/s 600 g

D 20 m/s 700 g

E 30 m/s 400 g

F 40 m/s 600 g

G 30 m/s 600 g

H 20 m/s 500 g

Greatest 1_____ 2_____ 3_____ 4_____ 5_____ 6_____ 7_____ 8_____ Least

Or, all rockets have the same kinetic energy._____

Please carefully explain your reasoning.

How sure were you of your ranking? (circle one)

Basically Guessed				Sure					Very Sure
1	2	3	4	5	6	7	8	9	10

Sliding Masses on Incline—Kinetic Energy [64]

Rank, in order from greatest to least, the final kinetic energies of the sliding masses the instant before they reach the bottom of the incline. All surfaces are frictionless. All masses start from rest.

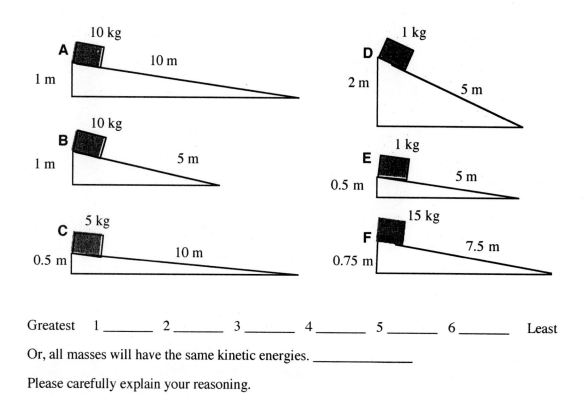

Greatest   1 _____ 2 _____ 3 _____ 4 _____ 5 _____ 6 _____   Least

Or, all masses will have the same kinetic energies. _____

Please carefully explain your reasoning.

How sure were you of your ranking? (circle one)
Basically Guessed   Sure   Very Sure
1   2   3   4   5   6   7   8   9   10

Sliding Masses on Incline—Change in Potential Energy [65]

Rank, in order from greatest to least, the change in gravitational potential energy of the sliding masses from the top of the incline to the bottom of the incline. All surfaces are frictionless. All masses start from rest at the top of the incline.

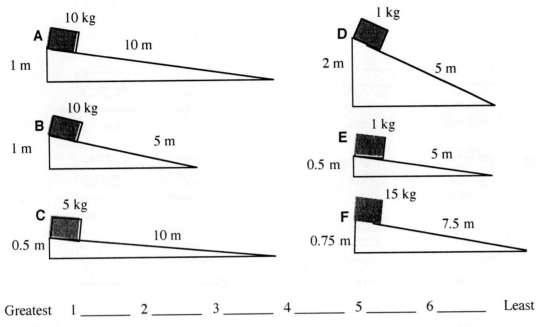

Greatest   1 _____ 2 _____ 3 _____ 4 _____ 5 _____ 6 _____ Least

Or, all masses will have the same change in gravitational potential energy. _____

Please carefully explain your reasoning.

How sure were you of your ranking? (circle one)
Basically Guessed   Sure   Very Sure
1   2   3   4   5   6   7   8   9   10

[65] P. Marquard, D. Maloney
Physics Ranking Tasks

Mechanics

Cars—Change in Kinetic Energy During a Change of Velocity [66]

The eight situations below show *before* and *after* "snapshots" of a car's velocity. Rank these situations, in terms of the change in kinetic energy of these cars, from most positive to most negative. All cars have the same mass and have traveled the same distance during this change. Negative numbers, if any, rank lower than positive ones (-20 m/s < -10 m/s < 0 < 5).

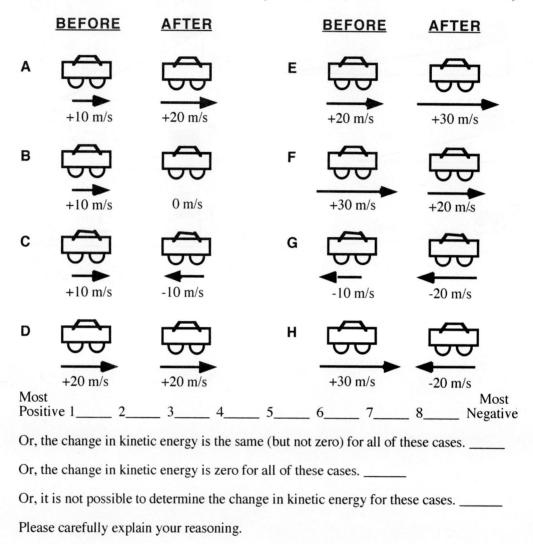

Most
Positive 1_____ 2_____ 3_____ 4_____ 5_____ 6_____ 7_____ 8_____ Most Negative

Or, the change in kinetic energy is the same (but not zero) for all of these cases. _____

Or, the change in kinetic energy is zero for all of these cases. _____

Or, it is not possible to determine the change in kinetic energy for these cases. _____

Please carefully explain your reasoning.

How sure were you of your ranking? (circle one)

Basically Guessed				Sure				Very Sure	
1	2	3	4	5	6	7	8	9	10

[66] J. Cole, D. Maloney, C. Hieggelke

The following drawings indicate the motion of a ball subject to one or more forces on various surfaces from left to right. Each circle represents the position of the ball at succeeding instants of time. Each time-interval between positions is equal. In all situations, the balls start with the same velocity.

Rank each case from the highest to the lowest final kinetic energy based on the figures using the coordinate system shown in the diagram. Assume the acceleration for each situation to be constant.

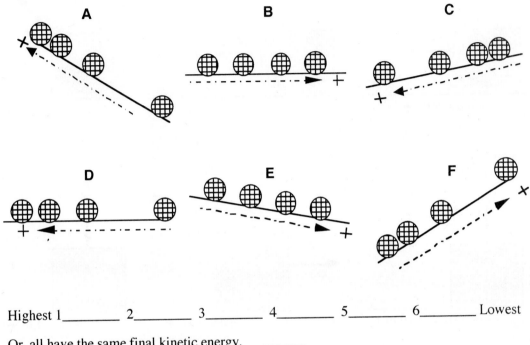

Highest 1_____ 2_____ 3_____ 4_____ 5_____ 6_____ Lowest

Or, all have the same final kinetic energy._____

Please carefully explain your reasoning.

How sure were you of your ranking? (circle one)

Basically Guessed   Sure   Very Sure

1   2   3   4   5   6   7   8   9   10

[67] D. Maloney
Physics Ranking Tasks

Mechanics

In the figures below, identical boxes of mass 10 kg are moving at the same initial velocity to the right on a flat surface. The same magnitude force, F, is applied to each box for the distance, d, indicated in the figures.

Rank these situations in order of the work done on the box by F while the box moves the indicated distance to the right.

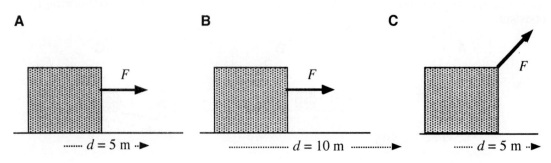

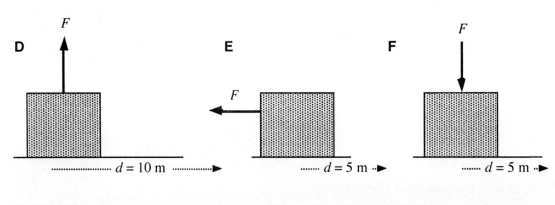

Greatest 1_____ 2_____ 3_____ 4_____ 5_____ 6_____ Least

Or, all of the boxes experience the same work. _____

Please carefully explain your reasoning.

How sure were you of your ranking? (circle one)

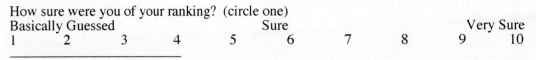

Equal Force on Boxes—Work Done on Hand [69]

In the figures below, identical boxes of mass 10 kg are moving at the same initial velocity to the right on a flat surface. The same magnitude force, F, is applied by a hand to each box for the distance, d, indicated in the figures.

Rank these situations in order of the work done by the box on the hand causing F while the box moves the indicated distance to the right.

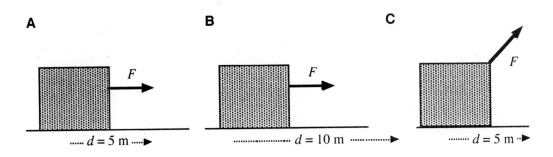

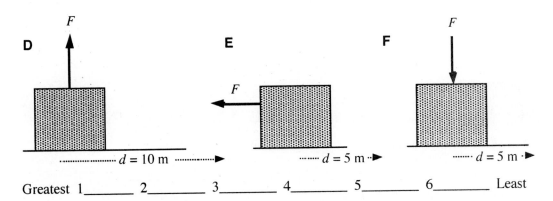

Greatest 1_____ 2_____ 3_____ 4_____ 5_____ 6_____ Least

Or, all of the boxes do the same (nonzero) work on the hand. _____

Or, all of the boxes do no work on the hand. _____

Or, it is not possible to determine the work done on the hand. _____

Please carefully explain your reasoning.

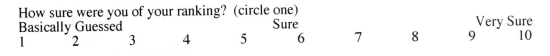

How sure were you of your ranking? (circle one)

Basically Guessed					Sure			Very Sure	
1	2	3	4	5	6	7	8	9	10

[69] E. Eckard, D. Maloney, C. Hieggelke

Velocity Time Graph—Work Done on Box [70]

Shown below is a graph of velocity versus time for an object that moves along a straight, horizontal line under the, perhaps intermittent, action of a single force exerted by an external agent.

Rank the intervals shown on the graph, from greatest to least, on the basis of the work done on the object by the external agent.

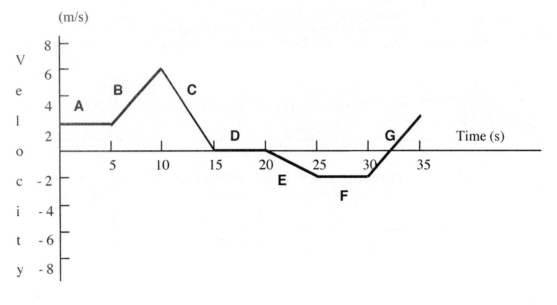

Greatest 1_____ 2_____ 3_____ 4_____ 5_____ 6_____7 _____ Least

Or, the agent does the same amount of work during all of these intervals. _____

Or, the agent does not work during any of these intervals. _____

Please carefully explain your reasoning.

How sure were you of your ranking? (circle one)

Basically Guessed				Sure				Very Sure	
1	2	3	4	5	6	7	8	9	10

[70] C. Hieggelke, D. Maloney, T. O'Kuma

Pendulums—Maximum Speed of the Bob[71]

Shown below are six situations where spheres are attached to strings forming pendulums. The pendulums vary in mass and length, but the angles from the vertical are the same for all.

Rank these situations, from greatest to least, on the basis of the maximum speed of the bob at the bottom of the swing. In other words, put first the pendulum whose bob has the greatest speed going through the equilibrium point and put last the pendulum whose bob has the least speed at equilibrium.

A

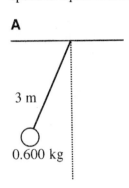

3 m

0.600 kg

B

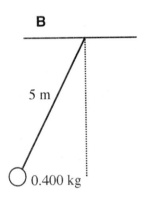

5 m

0.400 kg

C

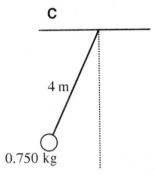

4 m

0.750 kg

D

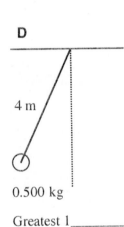

4 m

0.500 kg

E

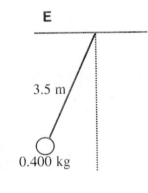

3.5 m

0.400 kg

F

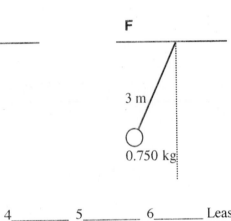

3 m

0.750 kg

Greatest 1_____ 2_____ 3_____ 4_____ 5_____ 6_____ Least

Or, the maximum speed is the same for all six of these. _____

Please carefully explain your reasoning.

How sure were you of your ranking? (circle one)

Basically Guessed   Sure   Very Sure

1   2   3   4   5   6   7   8   9   10

[71] C. Hieggelke, D. Maloney, T. O'Kuma
Physics Ranking Tasks

Mechanics

Force Pushing Box—Change in Kinetic Energy [72]

Various similar boxes are being pushed for 10 m across a floor by a *net* horizontal force as shown below. The mass of the boxes and the net horizontal force for each case are given in the indicated figures.

Rank the change in kinetic energy for each box from the greatest change in kinetic energy to the least change in kinetic energy. All boxes have an initial velocity of +10 m/s (+ direction is to the right and - to the left with –4 < -2).

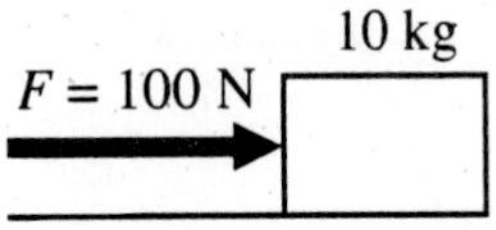

10 kg	15 kg	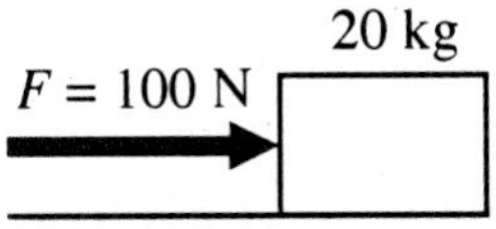 20 kg
$F = 100$ N	$F = 50$ N	$F = 75$ N
A	**B**	**C**

20 kg	10 kg	15 kg
$F = 50$ N	$F = 75$ N	$F = 100$ N
D	**E**	**F**

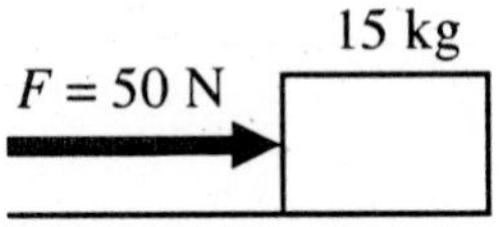

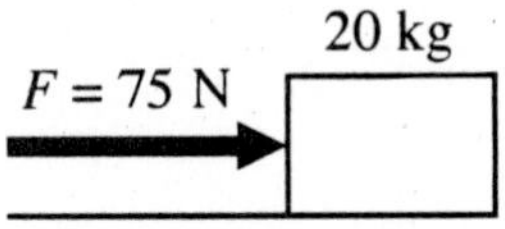

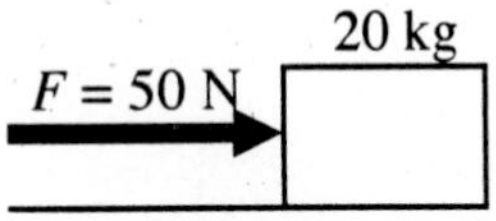

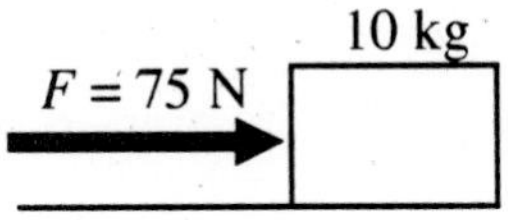

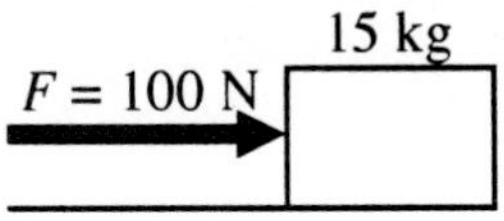

10 kg	20 kg
$F = 50$ N	$F = 100$ N
G	**H**

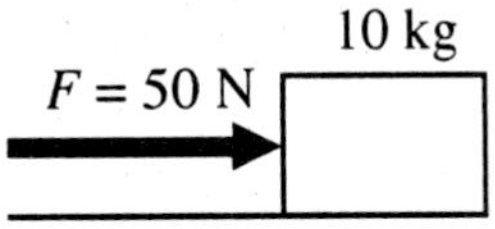

Greatest 1 _____ 2 _____ 3 _____ 4 _____ 5 _____ 6 _____ 7 _____ 8 _____ Least

Or, all changes in kinetic energy are the same._____

Please carefully explain your reasoning.

How sure were you of your ranking? (circle one)
Basically Guessed   Sure   Very Sure
1   2   3   4   5   6   7   8   9   10

Pushing Box with Friction—Change in Kinetic Energy [73]

Various similar boxes are being pushed for 10 m across a floor by a horizontal force as shown below. The weights of the boxes and the applied horizontal force for each case are given in the indicated figures. The frictional force is 20% of the weight of the box ($g = 10$ N/kg).

Rank the change in kinetic energy for each box from the greatest change in kinetic energy to the least change in kinetic energy. All boxes have an initial velocity of +10 m/s (+ direction is to the right and - to the left, with −4 < -2).

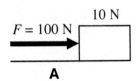

A

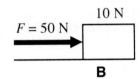

B

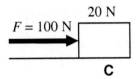

C

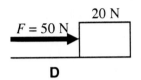

D

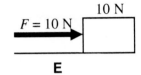

E

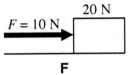

F

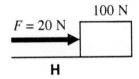

G

H

Greatest 1 _____ 2 _____ 3 _____ 4 _____ 5 _____ 6 _____ 7 _____ 8 _____ Least

Or, all changes in kinetic energy are the same._____

Please carefully explain your reasoning.

How sure were you of your ranking? (circle one)

Basically Guessed   Sure   Very Sure

1   2   3   4   5   6   7   8   9   10

Impulse—Momentum Ranking Tasks

The figures below depict carts moving along a horizontal surface at the speeds specified. The masses of the carts vary; specific values are given in the figures. The carts hit, compress the springs to some maximum amount, and then rebound. All of the spring systems are identical, exerting the same force on the carts, and the carts all hit the springs exactly the same way. The carts are not self-propelled, so they compress the springs for some maximum amount of time before stopping.

Rank the situations in order from the greatest time of compression to the least time of compression. That is, put first the cart that takes the longest time to reach maximum compression, and put last the cart that takes the shortest time.

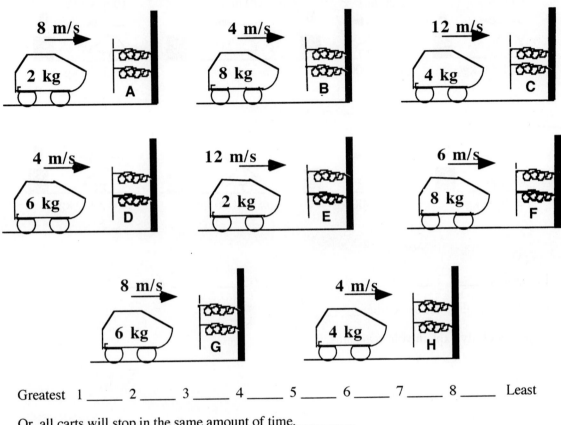

Greatest 1 _____ 2 _____ 3 _____ 4 _____ 5 _____ 6 _____ 7 _____ 8 _____ Least

Or, all carts will stop in the same amount of time. _____

Please carefully explain your reasoning.

How sure were you of your ranking?
Basically Guessed   Sure   Very Sure
1   2   3   4   5   6   7   8   9   10

Cars and Barriers—Stopping Time with the Same Force[75]

Shown below are eight cars that are moving along horizontal roads at specified speeds. Also given are the masses of the cars. All of the cars are the same size and shape, but they are carrying loads with different masses. All of these cars are going to be stopped by plowing into identical barriers. All of the cars are going to be stopped by the same constant force by the barrier.

Rank these situations from greatest to least on the basis of the stopping time that will be needed to stop the cars with the same force. That is, put first the car that requires the longest time and put last the car that requires the shortest time to stop the car with the same force.

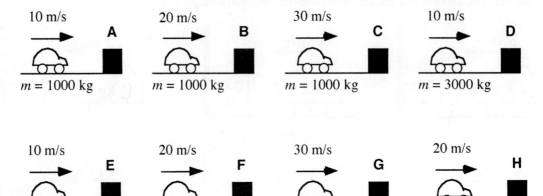

Longest   1 _____ 2 _____ 3 _____ 4 _____ 5 _____ 6 _____ 7 _____ 8 _____   Shortest

Or, all cars require the same time._____

Please carefully explain your reasoning.

How sure were you of your ranking? (circle one)
Basically Guessed   Sure   Very Sure
1   2   3   4   5   6   7   8   9   10

Exploding Shells—Final Location of Center of Mass[76]

Shown below are six situations where shells are at the top of their trajectories. These shells explode, at this instant, into two pieces with one piece, m_1, falling straight down (vertically) to the ground. All of these projectiles where fired from the same point at the same angle. We are told the speed of the shell at the top of the trajectory and the masses of the two pieces. Ignore air resistance in this situation.

Rank these situations, from greatest to least, on the basis of how far from the launch point the center of mass of the projectile ends up. That is, put first the situation where the final location of the center of mass is farthest from the launch point and put last the situation where it ends up closest to the launch point.

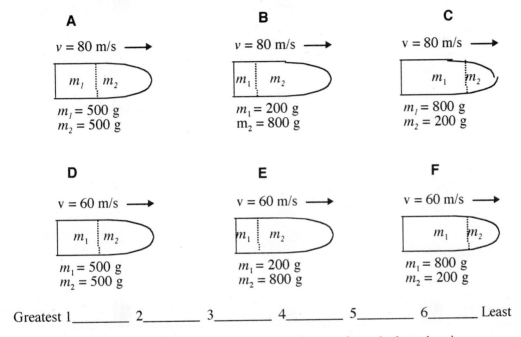

Greatest 1_____ 2_____ 3_____ 4_____ 5_____ 6_____ Least

Or, all six of the centers of mass end up the same distance from the launch point. _____

Please carefully explain your reasoning.

How sure were you of your ranking? (circle one)

Basically Guessed   Sure   Very Sure

1   2   3   4   5   6   7   8   9   10

[76] D. Maloney
Physics Ranking Tasks

Mechanics

Bouncing Cart—Change in Momentum I[77]

A cart with a spring plunger runs into a fixed barrier. The mass of the cart, its velocity just before impact with the barrier, and its velocity right after collision are given in each figure.

Rank the change in momentum for each cart from the greatest change in momentum to the least change in momentum (+ direction is to the right and - to the left with −4 < -2).

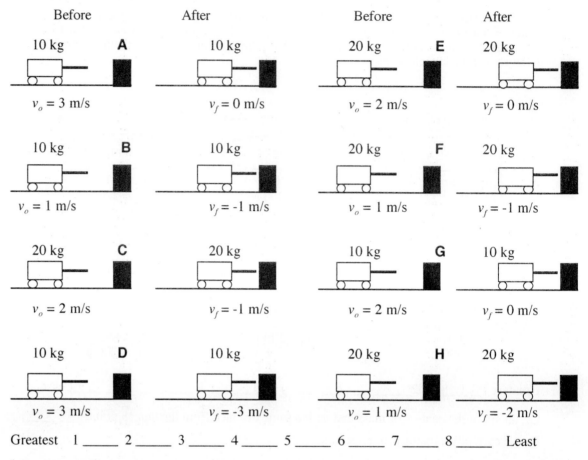

Before	After	Before	After
10 kg **A**	10 kg	20 kg **E**	20 kg
$v_o = 3$ m/s	$v_f = 0$ m/s	$v_o = 2$ m/s	$v_f = 0$ m/s
10 kg **B**	10 kg	20 kg **F**	20 kg
$v_o = 1$ m/s	$v_f = -1$ m/s	$v_o = 1$ m/s	$v_f = -1$ m/s
20 kg **C**	20 kg	10 kg **G**	10 kg
$v_o = 2$ m/s	$v_f = -1$ m/s	$v_o = 2$ m/s	$v_f = 0$ m/s
10 kg **D**	10 kg	20 kg **H**	20 kg
$v_o = 3$ m/s	$v_f = -3$ m/s	$v_o = 1$ m/s	$v_f = -2$ m/s

Greatest   1 _____ 2 _____ 3 _____ 4 _____ 5 _____ 6 _____ 7 _____ 8 _____   Least

Or, all the changes in momentum are the same._____

Please carefully explain your reasoning.

How sure were you of your ranking? (circle one)

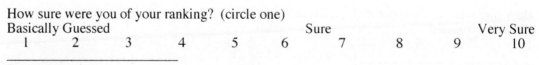

Basically Guessed   Sure   Very Sure
    1   2   3   4   5   6   7   8   9   10

[77] T. O'Kuma

Bouncing Cart—Change in Momentum II[78]

A cart with a spring plunger runs into a fixed barrier. The mass of the cart, its velocity just before impact with the barrier, and its velocity right after collision are given in each figure.

Rank the change in momentum for each cart from the greatest change in momentum to the least change in momentum (+ direction is to the right and - to the left with −4 < -2).

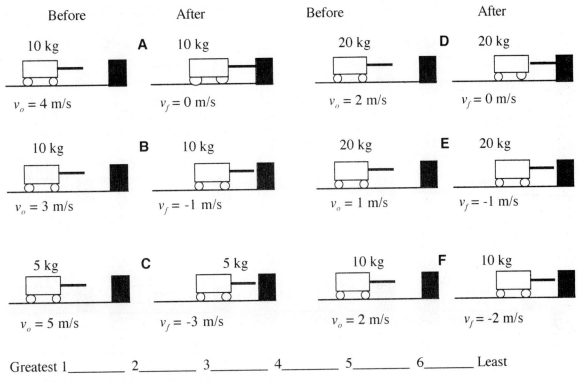

Before	After	Before	After

A 10 kg / 10 kg: $v_o = 4$ m/s → $v_f = 0$ m/s

D 20 kg / 20 kg: $v_o = 2$ m/s → $v_f = 0$ m/s

B 10 kg / 10 kg: $v_o = 3$ m/s → $v_f = -1$ m/s

E 20 kg / 20 kg: $v_o = 1$ m/s → $v_f = -1$ m/s

C 5 kg / 5 kg: $v_o = 5$ m/s → $v_f = -3$ m/s

F 10 kg / 10 kg: $v_o = 2$ m/s → $v_f = -2$ m/s

Greatest 1_____ 2_____ 3_____ 4_____ 5_____ 6_____ Least

Or, all the changes in momentum are the same._____

Please carefully explain your reasoning.

How sure were you of your ranking? (circle one)

Basically Guessed   Sure   Very Sure

1   2   3   4   5   6   7   8   9   10

A cart with a spring plunger runs into a fixed barrier. The mass of the cart, its velocity just before impact with the barrier, and its velocity right after collision are given in each figure.

Rank the change in momentum for each cart from the greatest change in momentum to the least change in momentum (+ direction is to the right and - to the left with –4 < -2).

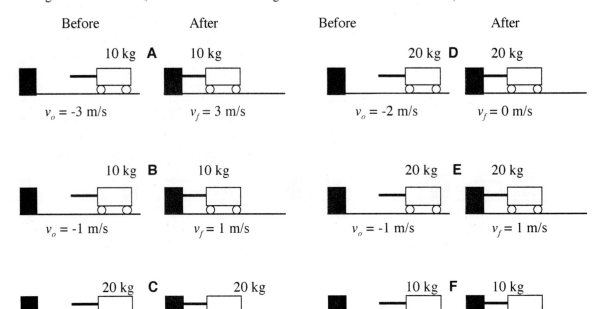

Greatest 1_____ 2_____ 3_____ 4_____ 5_____ 6_____ Least

Or, all the changes in momentum are the same._____

Please carefully explain your reasoning.

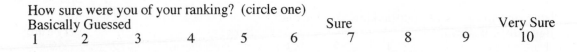

How sure were you of your ranking? (circle one)

Basically Guessed   Sure   Very Sure

1   2   3   4   5   6   7   8   9   10

Force Pushing Box—Change in Momentum[80]

Various similar boxes are being pushed for 10 seconds across a floor by a net horizontal force as shown below. The mass of the boxes and the net horizontal force for each case are given in the indicated figures.

Rank the change in momentum for each box from the greatest change in momentum to the least change in momentum. All boxes have an initial velocity of 0 m/s (+ direction is to the right and - to the left with −4 < -2).

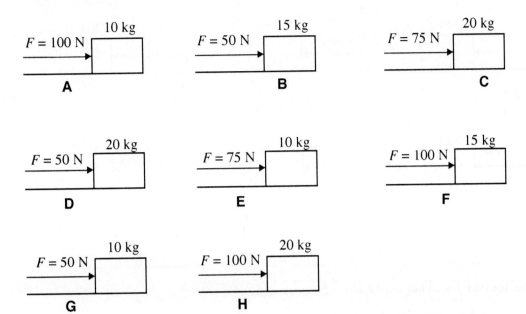

Greatest 1 _____ 2 _____ 3 _____ 4 _____ 5 _____ 6 _____ 7 _____ 8 _____ Least

Or, all the changes in momentum are the same._____

Please carefully explain your reasoning.

Force Pushing Box—Final Momentum[81]

Various similar boxes are being pushed for 10 seconds across a floor by a net horizontal force as shown below. The mass of the boxes and the net horizontal force for each case is given in the indicated figures.

Rank the final momentum for each box from the greatest momentum to the least momentum. All boxes have an initial velocity of 10 m/s. Motion to the right is positive.

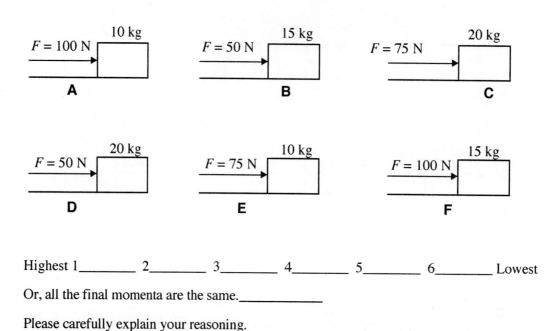

Highest 1_____ 2_____ 3_____ 4_____ 5_____ 6_____ Lowest

Or, all the final momenta are the same._____

Please carefully explain your reasoning.

How sure were you of your ranking? (circle one)

Basically Guessed   Sure   Very Sure

1   2   3   4   5   6   7   8   9   10

Cars—Impulse During a Change of Velocity [82]

The eight situations below show *before* and *after* "snapshots" of a car's velocity. Rank these situations, in terms of impulse on these cars, from most positive to most negative, to create these changes in velocity. All cars have the same mass. Negative numbers, if any, rank lower than positive ones (-20 m/s < -10 m/s < 0 < 5).

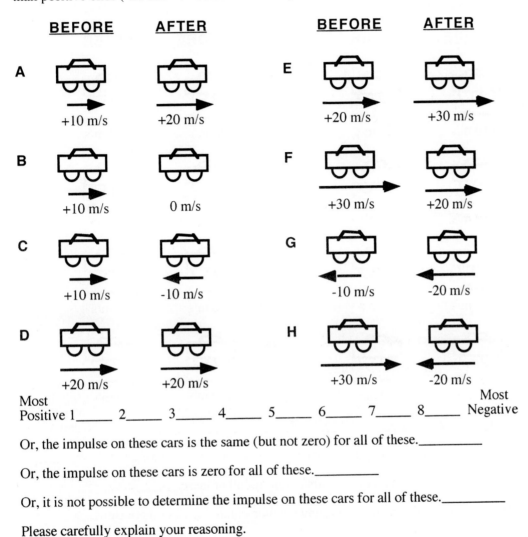

Most
Positive 1_____ 2_____ 3_____ 4_____ 5_____ 6_____ 7_____ 8_____ Most Negative

Or, the impulse on these cars is the same (but not zero) for all of these._____

Or, the impulse on these cars is zero for all of these._____

Or, it is not possible to determine the impulse on these cars for all of these._____

Please carefully explain your reasoning.

How sure were you of your ranking? (circle one)

Basically Guessed					Sure			Very Sure	
1	2	3	4	5	6	7	8	9	10

[82] J. Cole, D. Maloney, C. Hieggelke
Physics Ranking Tasks

Mechanics

The eight situations below show *before* and *after* "snapshots" of a car's velocity. Rank these situations, in terms of the change in momentum of these cars, from most positive to most negative. All cars have the same mass. Negative numbers, if any, rank lower than positive ones (-20 m/s < -10 m/s < 0< 5).

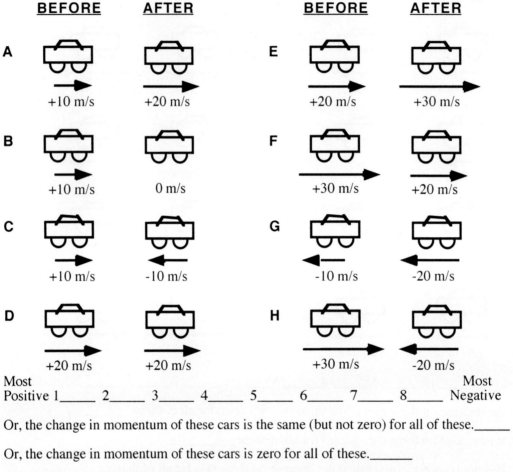

Most
Positive 1_____ 2_____ 3_____ 4_____ 5_____ 6_____ 7_____ 8_____ Negative

Or, the change in momentum of these cars is the same (but not zero) for all of these._____

Or, the change in momentum of these cars is zero for all of these._____

Or, it is not possible to determine the change in momentum for all of these cases._____

Please carefully explain your reasoning.

How sure were you of your ranking? (circle one)
Basically Guessed   Sure   Very Sure
1   2   3   4   5   6   7   8   9   10

Rotation Ranking Tasks

Four L's Rotating About an Axis (side view)—Moment of Inertia[84]

Below are four identical figure **L**'s, which are constructed from two rods of equal lengths and masses. For each figure, a different axis of rotation is indicated by the small circle with the dot inside, which indicates an axis that is perpendicular to the plane of the **L**'s. The axis of rotation is located either at the center or one end of a rod for each figure.

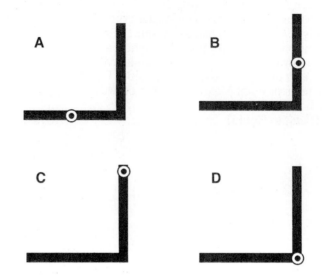

Rank these **L** figures according to their moments of inertia about the indicated axes, from largest to smallest. Ignore the width of each rod but not the length.

Largest   1 _____   2 _____   3 _____   4 _____   Smallest

Or, all these **L** systems have the same moment of inertia. _____

Please carefully explain your reasoning.

How sure were you of your ranking? (circle one)
Basically Guessed   Sure   Very Sure
1   2   3   4   5   6   7   8   9   10

[84] C. Hieggelke

Five T's Rotating About an Axis (top view)—Moment of Inertia [85]

Below are five identical figure **T**'s, which are constructed from two rods of equal lengths and masses. For each figure, a different axis of rotation in the plane of the paper is indicated by the dotted line. The axis of rotation is located either at the center or one end of a rod for each figure.

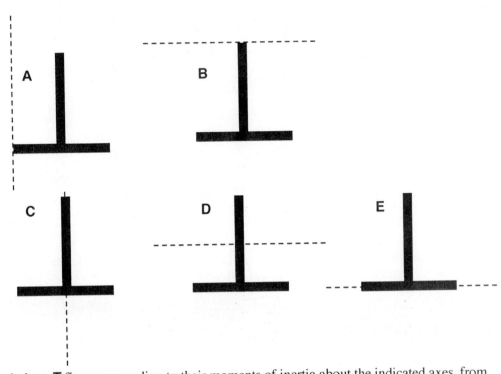

Rank these **T** figures according to their moments of inertia about the indicated axes, from largest to smallest. Ignore the width of each rod but not the length.

Largest   1 _____   2 _____   3 _____   4 _____   5 _____   Smallest

Or, all these **T** systems have the same moment of inertia. _____

Please carefully explain your reasoning.

How sure were you of your ranking? (circle one)

Basically Guessed   Sure   Very Sure

1   2   3   4   5   6   7   8   9   10

Four T's Rotating About an Axis (side view)—Moment of Inertia [86]

Below are four identical figure **T**'s, which are constructed from two rods of equal lengths and masses. For each figure, a different axis of rotation is indicated by the small circle with the dot inside, which indicates an axis that is perpendicular to the plane of the **T**'s. The axis of rotation is located either at the center or one end of a rod for each figure.

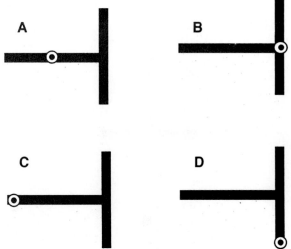

Rank these **T** figures according to their moments of inertia about the indicated axes, from largest to smallest. Ignore the width of each rod but not the length.

Largest   1 _____   2 _____   3 _____   4 _____   Smallest

Or, all these **T** systems have the same moment of inertia._____

Please carefully explain your reasoning.

Five T's Rotating About an Axis (top view)—Net Gravitational Torque[87]

Shown below (in a top view with the gravitation force into the page ⊕) are five identical figure **T**'s, which are constructed from two rods of equal lengths and masses. For each figure, a different axis of rotation, which is in the plane of the page, is indicated by the dotted line. The axes are either at the center or one end of a rod.

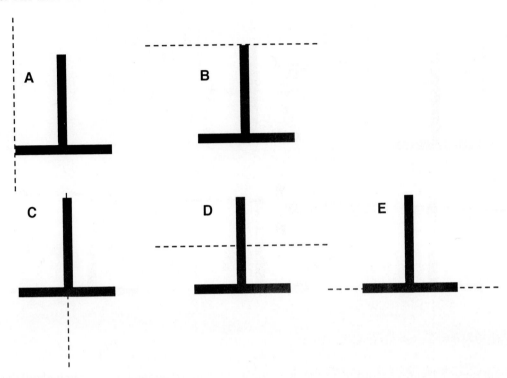

Rank these **T** figures according to the magnitudes of the net torque about the indicated axes, from largest to smallest.

Largest   1 _____   2 _____   3 _____   4 _____   5 _____   Smallest

Or, all these **T** systems have the same magnitude of the net torque. _____

Please carefully explain your reasoning.

How sure were you of your ranking? (circle one)
Basically Guessed   Sure   Very Sure
1   2   3   4   5   6   7   8   9   10

[87] C. Hieggelke
Physics Ranking Tasks

Mechanics

Five T's Rotating About an Axis (top view)—Angular Acceleration[88]

Shown below (in a top view with the gravitation force into the page $\oplus$) are five identical figure **T**'s, which are constructed from two rods of equal lengths and masses. For each figure, a different axis of rotation, which is in the plane of the page, is indicated by the dotted line. The axes are either at the center or one end of a rod.

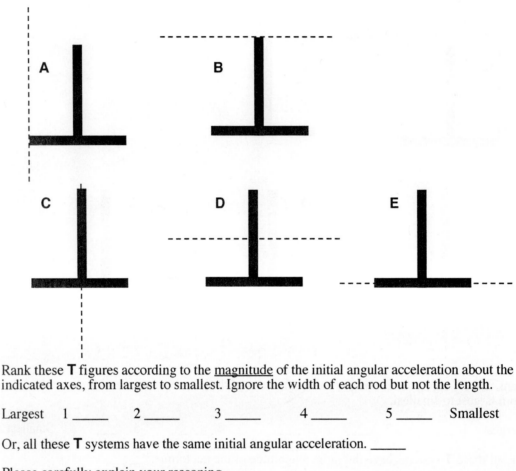

Rank these **T** figures according to the <u>magnitude</u> of the initial angular acceleration about the indicated axes, from largest to smallest. Ignore the width of each rod but not the length.

Largest   1 _____   2 _____   3 _____   4 _____   5 _____   Smallest

Or, all these **T** systems have the same initial angular acceleration. _____

Please carefully explain your reasoning.

How sure were you of your ranking? (circle one)
Basically Guessed   Sure   Very Sure
1   2   3   4   5   6   7   8   9   10

Blocks on Rotating Turntables—Horizontal Force[89]

Shown below in a top view are six blocks that are sitting at rest on rotating turntables. All of the turntables have the same rotation rate. The masses of the blocks and how far out from the center they sit varies. Specific values of the variable are given in the figures.

Rank these blocks, from greatest to least, on the basis of the magnitude of the horizontal forces holding the blocks on the turntables. That is, put first the block that has the largest force holding it on the turntable and put last the block that has the weakest force holding it on the turntable.

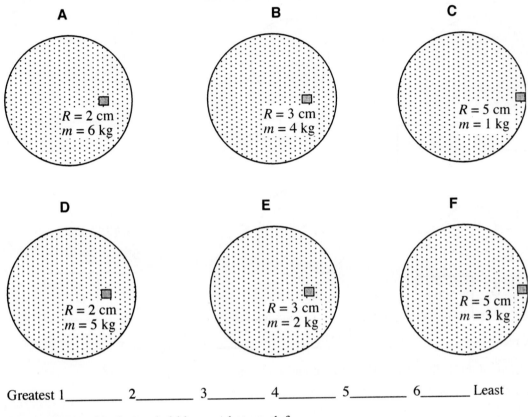

A
$R = 2$ cm
$m = 6$ kg

B
$R = 3$ cm
$m = 4$ kg

C
$R = 5$ cm
$m = 1$ kg

D
$R = 2$ cm
$m = 5$ kg

E
$R = 3$ cm
$m = 2$ kg

F
$R = 5$ cm
$m = 3$ kg

Greatest 1_____ 2_____ 3_____ 4_____ 5_____ 6_____ Least

Or, all of these blocks are held by equal strength forces. _____

Please carefully explain your reasoning.

How sure were you of your ranking? (circle one)

Basically Guessed				Sure				Very Sure	
1	2	3	4	5	6	7	8	9	10

[89] C. Hieggelke, D. Maloney, T. O'Kuma
Physics Ranking Tasks

Mechanics

Hanging Weights and Fixed Disks—Torque[90]

Shown below are six situations where vertically oriented circular disks have strings wrapped around them. The other ends of the strings are attached to hanging masses. The radii of the disks, the masses of the disks, and the masses of the hanging masses all vary. The disks are fixed and are *not* free to rotate. Specific values of the variables are given in the figures.

Rank these situations, from greatest to least, on the basis of the magnitude of the torque on the disks. That is, put first the situation where the disk has the greatest torque acting on it and put last the situation where the disk has the least torque acting on it.

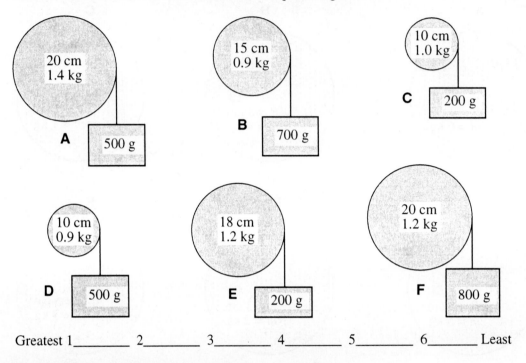

Greatest 1_____ 2_____ 3_____ 4_____ 5_____ 6_____ Least

Or, all of these disks have the same torque acting on them. _____

Please carefully explain your reasoning.

How sure were you of your ranking? (circle one)

Basically Guessed				Sure				Very Sure	
1	2	3	4	5	6	7	8	9	10

[90] C. Hieggelke, D. Maloney, T. O'Kuma

Shown below in a top view are six uniform rods that vary in mass (*M*) and length (*L*). Also shown are circles representing a vertical axis around which the rods are going to be rotated in a horizontal plane and arrows representing forces acting to rotate the rods. The forces change direction in order to always act perpendicular to the rods. Specific values for the lengths and masses of the rods and the magnitudes of the forces are given in each figure.

Rank these rods, from greatest to least, on the basis of the magnitude of their angular acceleration. That is, put first the rod that has the largest angular acceleration and put last the one that will have the smallest angular acceleration.

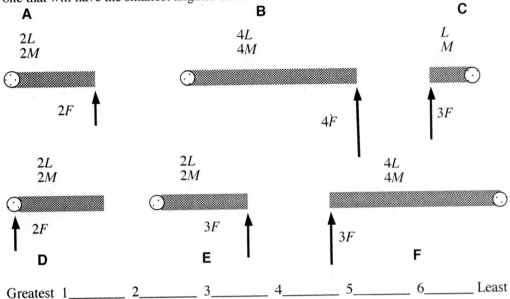

Greatest 1_____ 2_____ 3_____ 4_____ 5_____ 6_____ Least

Or, all six of these rods will have the same magnitude angular acceleration. _____

Please carefully explain your reasoning.

How sure were you of your ranking? (circle one)

Basically Guessed					Sure			Very Sure	
1	2	3	4	5	6	7	8	9	10

[91] C. Hieggelke, D. Maloney, T. O'Kuma
Physics Ranking Tasks

Mechanics

Horizontal Uniform Rods—Change in Angular Momentum[92]

Shown below in a top view are six uniform rods that vary in mass (*M*) and length (*L*). Also shown are circles representing a vertical axis around which the rods are going to be rotated in a horizontal plane and arrows representing forces acting to rotate the rods. The forces change direction in order to always act perpendicular to the rods. Specific values for the lengths and masses of the rods and the magnitudes of the forces are given in each figure.

Rank these rods, from greatest to least, on the basis of their change in the magnitude of angular momentum for the same time period. That is, put first the rod that has the largest change in angular momentum and put last the one that will have the smallest change.

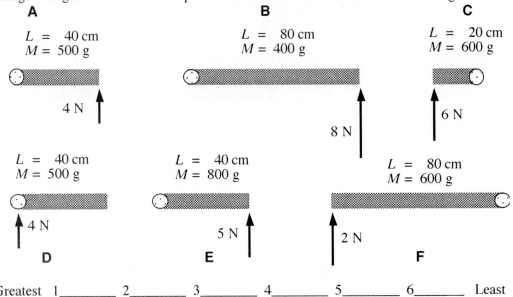

Greatest 1_____ 2_____ 3_____ 4_____ 5_____ 6_____ Least

Or, all six of these rods will have the same change in magnitude of angular momentum. _____

Please carefully explain your reasoning.

How sure were you of your ranking? (circle one)
Basically Guessed   Sure   Very Sure
1   2   3   4   5   6   7   8   9   10

Rotating Systems of Point Masses—Difficult to Rotate[93]

Shown below are six arrangements of 10-point masses. Each of the point masses is the same size and has the same mass. Also shown in each figure is a solid line representing an axis about which the masses are going to be rotated. The point masses exert forces on each other so that they all maintain the arrangements shown while being rotated.

Rank these arrangements, from greatest to least, on the basis of how hard it will be to start the arrangements rotating. That is, put first the arrangement that will be the most difficult to start rotating and put last the easiest arrangement.

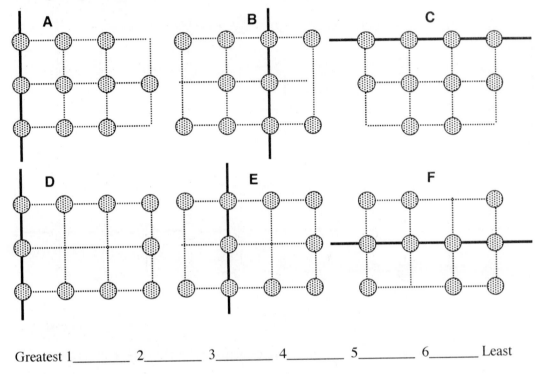

Greatest 1_____ 2_____ 3_____ 4_____ 5_____ 6_____ Least

Or, all of these arrangements will be equally difficult to rotate. _____

Please carefully explain your reasoning.

How sure were you of your ranking? (circle one)

Basically Guessed   Sure   Very Sure

1   2   3   4   5   6   7   8   9   10

[93] C. Hieggelke, D. Maloney, T. O'Kuma

Physics Ranking Tasks

Mechanics

Rotating Systems of Point Masses—Center of Mass[94]

Shown below are six arrangements of 10-point masses. Each of the point masses is the same size and has the same mass. Also shown in each figure is a solid line representing an axis about which the masses are going to be rotated. The point masses exert forces on each other so that they all maintain the arrangements shown while being rotated.

Rank these arrangements, from greatest to least, on the basis of the distance between the center of mass and the axis of rotation.

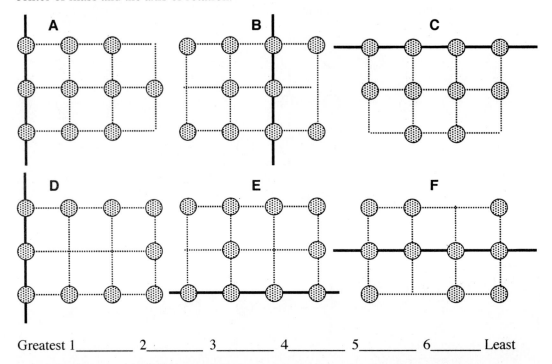

Greatest 1_____ 2_____ 3_____ 4_____ 5_____ 6_____ Least

Or, all of these arrangements have the same distance between the center of mass and axis of rotation . _____

Please carefully explain your reasoning.

How sure were you of your ranking? (circle one)
Basically Guessed   Sure   Very Sure
1   2   3   4   5   6   7   8   9   10

[94] C. Hieggelke, D. Maloney, T. O'Kuma

Shown below are seven situations where a student is holding a meter stick at the left end at various angles. A 1000 g mass is hung on the meter sticks at different locations. All of the meter sticks are identical, but the distance along the meter stick at which the 1000 g mass is hung and the angles at which the student holds the meter stick vary. Specific values are given in each figure. (Ignore the mass of the meter stick.)

Rank these situations, from greatest to least, on the basis of how difficult it would be for the student to hold the meter stick from the left end in the position shown. That is, put first the situation where it would be hardest to hold the meter stick at the angle shown and put last the situation where it would be easiest to hold it at the angle shown.

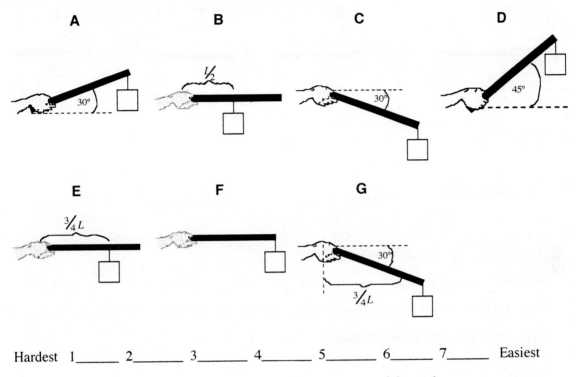

Hardest 1_____ 2_____ 3_____ 4_____ 5_____ 6_____ 7_____ Easiest

Or, it would be equally difficult to hold all seven of these meter sticks as shown. _____

Please carefully explain your reasoning.

How sure were you of your ranking? (circle one)
Basically Guessed   Sure   Very Sure
1 2 3 4 5 6 7 8 9 10

Shown below are seven situations where a student is holding a meter stick straight out horizontally. All of the meter sticks are identical, but the number, mass, and locations (either at the 50 cm mark or at the other end of the meter stick from the student) of the objects hung on the sticks vary. The specific values and locations are given in each figure.

Rank these situations, from greatest to least, on the basis of how difficult it would be for the student to keep the meter stick from rotating. That is, put first the situation where it would be hardest to hold the meter stick horizontal and put last the situation where it would be easiest to hold it horizontal.

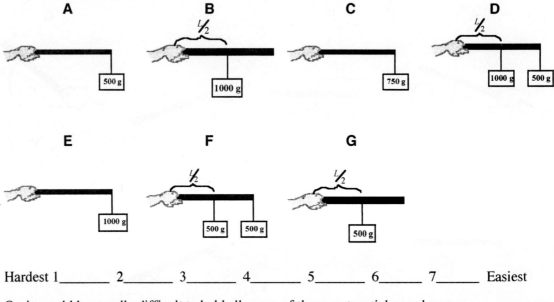

Hardest 1_____ 2_____ 3_____ 4_____ 5_____ 6_____ 7_____ Easiest

Or, it would be equally difficult to hold all seven of these meter sticks as shown. _____

Please carefully explain your reasoning.

How sure were you of your ranking? (circle one)
Basically Guessed   Sure   Very Sure
1   2   3   4   5   6   7   8   9   10

[96] C. Hieggelke, D. Maloney, T. O'Kuma

Properties of Matter Ranking Tasks

Each figure below shows a block attached to the end of a spring resting on a frictionless surface. In each figure, the springs are to be stretched to the right by a distance given in the figure and released. The blocks will then proceed to oscillate. The mass is given for each different block, and the force constant is given for each different spring.

Rank the figures from greatest to least on the basis of the period of the vibratory motion. That is, rank the figures on the basis of how long each will take to go through one complete cycle.

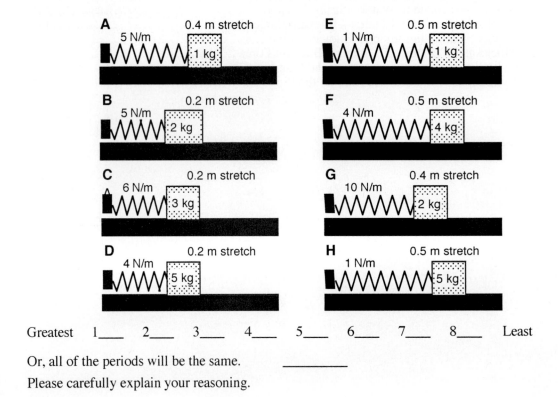

Greatest   1____   2____   3____   4____   5____   6____   7____   8____   Least

Or, all of the periods will be the same.   _____

Please carefully explain your reasoning.

How sure were you of your ranking? (circle one)

Basically Guessed				Sure				Very Sure	
1	2	3	4	5	6	7	8	9	10

Blocks Suspended in Liquids—Buoyant Force [98]

Shown below are six containers that contain various liquids. Blocks of various solids are suspended in the liquids by being hung from a supporting rod. All of these blocks are the same size, but they have different masses (labeled as M_b) since they are made of different materials. All of the containers have the same volume of liquid, but the masses of these liquids vary (labeled M_l) since the liquids are different. Specific values for the masses of the blocks and the liquids are given in each figure. The volume of the blocks is one-fifth the volume of the liquids.

Rank these situations, from greatest to least, on the basis of the buoyant forces on the blocks. That is, put first the block that experiences the greatest buoyant force, and put last the block that experiences the smallest buoyant force.

A

$M_b = \quad 40$ g
$M_l = \quad 200$ g

B

$M_b = \quad 50$ g
$M_l = \quad 200$ g

C

$M_b = \quad 30$ g
$M_l = \quad 150$ g

D

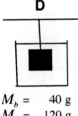

$M_b = \quad 40$ g
$M_l = \quad 120$ g

E

$M_b = \quad 20$ g
$M_l = \quad 80$ g

F

$M_b = \quad 30$ g
$M_l = \quad 120$ g

Greatest Force   1____   2____   3____   4____   5____   6____   Least Force

Or, all of the blocks experience the same buoyant force. _____

Or, there is no buoyant force on any of these blocks. _____

Please carefully explain your reasoning.

How sure were you of your ranking? (circle one)
Basically Guessed   Sure   Very Sure
1   2   3   4   5   6   7   8   9   10

[98] D. Maloney
Ranking Task Exercises in Physics   105   Properties of Matter, Heat-Thermodynamics, Waves

Shown below are six containers that contain various liquids. Blocks of various solids are suspended in the liquids by being hung from a supporting rod. All of these blocks are the same size, but they have different masses (labeled as M_b) since they are made of different materials. All of the containers have the same volume of liquid, but the masses of these liquids vary (labeled M_l) since the liquids are different. Specific values for the masses of the blocks and the liquids are given in each figure. The volume of the blocks is one-fifth the volume of the liquids.

Rank these situations, from greatest to least, on the basis of the volume of the liquid displaced by the blocks.

A

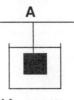

$M_b = \quad 40$ g
$M_l = \quad 200$ g

B

$M_b = \quad 50$ g
$M_l = \quad 200$ g

C

$M_b = \quad 30$ g
$M_l = \quad 150$ g

D

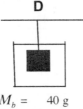

$M_b = \quad 40$ g
$M_l = \quad 120$ g

E

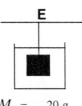

$M_b = \quad 20$ g
$M_l = \quad 80$ g

F

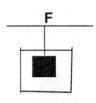

$M_b = \quad 30$ g
$M_l = \quad 120$ g

Greatest Volume   1____ 2____ 3____ 4____ 5____ 6____   Least Volume

Or, all of the volumes of the liquids displaced by the blocks are the same. _____

Please carefully explain your reasoning.

How sure were you of your ranking? (circle one)
Basically Guessed   Sure   Very Sure
1   2   3   4   5   6   7   8   9   10

Shown below are six containers that contain various liquids. Blocks of various solids are suspended in the liquids by being hung from a supporting rod. All of these blocks are the same size, but they have different masses (labeled as M_b) since they are made of different materials. All of the containers have the same volume of liquid, but the masses of these liquids vary (labeled M_l) since the liquids are different. Specific values for the masses of the blocks and the liquids are given in each figure. The volume of the blocks is one-fifth the volume of the liquids.

Rank these situations, from greatest to least, on the basis of the mass of the liquids displaced by the blocks.

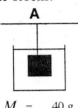

A

M_b = 40 g
M_l = 200 g

B

M_b = 50 g
M_l = 200 g

C

M_b = 30 g
M_l = 150 g

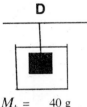

D

M_b = 40 g
M_l = 120 g

E

M_b = 20 g
M_l = 80 g

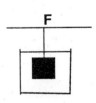

F

M_b = 30 g
M_l = 120 g

Greatest Mass   1____ 2____ 3____ 4____ 5____ 6____   Least Mass

Or, all of the masses of the liquids displaced by the blocks are the same. _____

Please carefully explain your reasoning.

How sure were you of your ranking? (circle one)
Basically Guessed   Sure   Very Sure
1   2   3   4   5   6   7   8   9   10

Shown below are six objects that have different masses and different volumes. These blocks are suspended at two different depths in water by being hung by a string from a supporting rod.

Rank these situations, from greatest to least, on the basis of buoyant force on the blocks by the water.

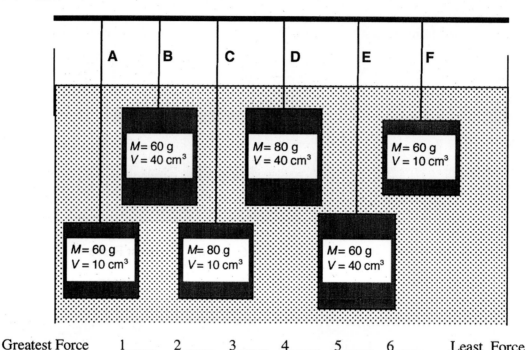

Greatest Force   1____   2____   3____   4____   5____   6____   Least Force

Or, all of the buoyant forces on the blocks by the water are equal. _____

Or, there are no buoyant forces on the blocks by the water. _____

Or, it is not possible to determine the buoyant forces on the blocks by the water. _____

Please carefully explain your reasoning.

How sure were you of your ranking? (circle one)

Basically Guessed				Sure				Very Sure	
1	2	3	4	5	6	7	8	9	10

Shown below are seven wood blocks which have different masses and different volumes. These blocks are floating in water. On top of these blocks are additional masses which provide a load for each of these blocks. Note: The blocks are not drawn correctly in terms of the depth the wooden blocks are in the water.

Rank these situations, from greatest to least, on the basis of buoyant force on the wood blocks by the water.

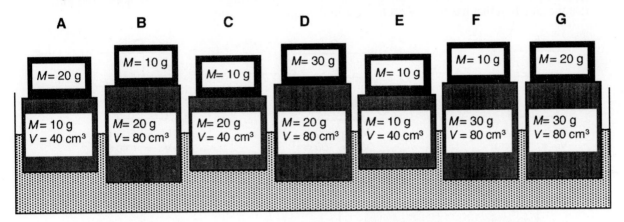

Greatest Force 1_____ 2_____ 3_____ 4_____ 5_____ 6_____ 7_____ Least Force

Or, all of the buoyant forces on the wood blocks by the water are equal. _____

Or, there are no buoyant forces on the wood blocks by the water. _____

Or, it is not possible to determine the buoyant forces on the wood blocks by the water. ___

Please carefully explain your reasoning.

How sure were you of your ranking? (circle one)

Basically Guessed				Sure				Very Sure	
1	2	3	4	5	6	7	8	9	10

Shown below are six containers that have the same volume of water in them. Blocks of various solids are suspended in the containers by being hung from a supporting rod. The blocks vary in both size and mass. The blocks are made of different materials, but all of the blocks would sink if the strings were cut. Specific values for the masses labeled as M_b and volumes labeled as V_b of the blocks are given in each figure.

Rank these situations, from greatest to least, on the basis of the tensions in the strings. That is, put first the situation that has the greatest tension in the string supporting the block, and put last the situation that has the lowest tension in the supporting string.

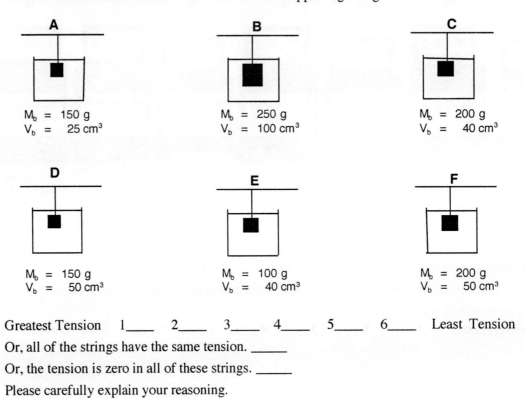

A
M_b = 150 g
V_b = 25 cm³

B
M_b = 250 g
V_b = 100 cm³

C
M_b = 200 g
V_b = 40 cm³

D
M_b = 150 g
V_b = 50 cm³

E
M_b = 100 g
V_b = 40 cm³

F
M_b = 200 g
V_b = 50 cm³

Greatest Tension   1____   2____   3____   4____   5____   6____   Least Tension

Or, all of the strings have the same tension. _____

Or, the tension is zero in all of these strings. _____

Please carefully explain your reasoning.

How sure were you of your ranking? (circle one)
Basically Guessed   Sure   Very Sure
1   2   3   4   5   6   7   8   9   10

Blocks Suspended in Liquids—Buoyant Force II

Shown below are six containers that have the same volume of the same liquid in them. Blocks of various solids are suspended in the containers by being hung from a supporting rod. The blocks vary in both size and mass. The blocks are made of different materials, but all of the blocks would sink if the strings were cut. Specific values for the masses labeled as M_b and volumes labeled as V_b of the blocks are given in each figure.

Rank these situations, from greatest to least, on the basis of the buoyant force on the blocks. That is, put first the situation that has the greatest buoyant force on the block, and put last the situation that has the lowest buoyant force on the block.

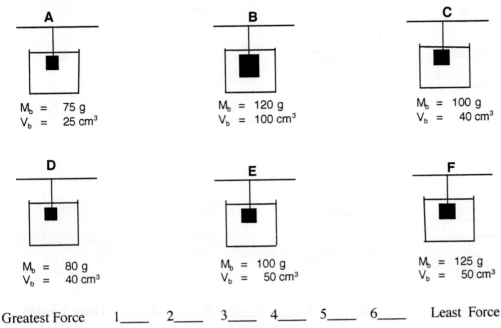

A
M_b = 75 g
V_b = 25 cm^3

B
M_b = 120 g
V_b = 100 cm^3

C
M_b = 100 g
V_b = 40 cm^3

D
M_b = 80 g
V_b = 40 cm^3

E
M_b = 100 g
V_b = 50 cm^3

F
M_b = 125 g
V_b = 50 cm^3

Greatest Force   1____ 2____ 3____ 4____ 5____ 6____   Least Force

Or, all of the blocks have the same nonzero buoyant force. _____

Or, the buoyant force is zero on all of these blocks. _____

Please carefully explain your reasoning.

How sure were you of your ranking? (circle one)
Basically Guessed   Sure   Very Sure
1   2   3   4   5   6   7   8   9   10

Shown below are eight containers that have the same volume of the same liquid in them. Blocks of various solids are floating on top of the liquid. The blocks vary in both size and mass. Specific values for the masses labeled as M_b and volumes labeled as V_b of the blocks are given in each figure.

Rank these situations, from greatest to least, on the basis of the buoyant force by the liquid on the blocks. That is, put first the situation that has the greatest buoyant force by the liquid on the block, and put last the situation that has the lowest buoyant force by the liquid on the block.

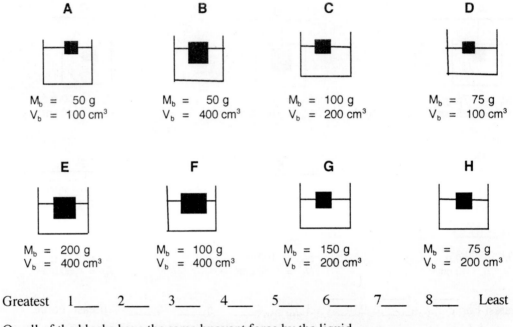

A	**B**	**C**	**D**
$M_b = 50$ g	$M_b = 50$ g	$M_b = 100$ g	$M_b = 75$ g
$V_b = 100$ cm³	$V_b = 400$ cm³	$V_b = 200$ cm³	$V_b = 100$ cm³

E	**F**	**G**	**H**
$M_b = 200$ g	$M_b = 100$ g	$M_b = 150$ g	$M_b = 75$ g
$V_b = 400$ cm³	$V_b = 400$ cm³	$V_b = 200$ cm³	$V_b = 200$ cm³

Greatest   1___   2___   3___   4___   5___   6___   7___   8___   Least

Or, all of the blocks have the same buoyant force by the liquid. _____

Please carefully explain your reasoning.

How sure were you of your ranking? (circle one)

Basically Guessed					Sure				Very Sure
1	2	3	4	5	6	7	8	9	10

Blocks at the Bottom of Liquids—Buoyant Force [106]

Shown below are eight containers that have the same volume of the same liquid in them. Blocks of various solids are at the bottom of the containers. The blocks vary in both size and mass. Specific values for the masses labeled as M_b and volumes labeled as V_b of the blocks are given in each figure.

Rank these situations, from greatest to least, on the basis of buoyant force on the blocks. That is, put first the situation that has the greatest buoyant force on the block, and put last the situation that has the lowest buoyant force on the block.

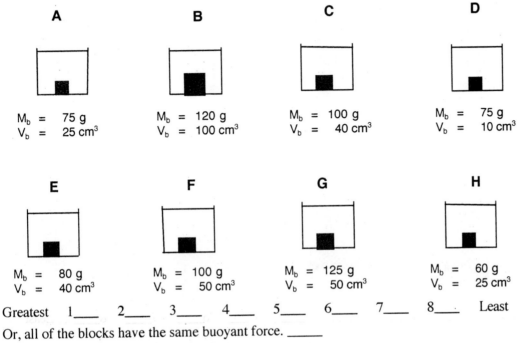

A

M_b = 75 g
V_b = 25 cm^3

B

M_b = 120 g
V_b = 100 cm^3

C

M_b = 100 g
V_b = 40 cm^3

D

M_b = 75 g
V_b = 10 cm^3

E

M_b = 80 g
V_b = 40 cm^3

F

M_b = 100 g
V_b = 50 cm^3

G

M_b = 125 g
V_b = 50 cm^3

H

M_b = 60 g
V_b = 25 cm^3

Greatest   1____   2____   3____   4____   5____   6____   7____   8____   Least

Or, all of the blocks have the same buoyant force. _____

Or, the buoyant force is zero on all these blocks. _____

Please carefully explain your reasoning.

How sure were you of your ranking? (circle one)
Basically Guessed   Sure   Very Sure
1   2   3   4   5   6   7   8   9   10

Cylinders of Water—Pressure on the Plug I [107]

The figures below show side views of eight hollow cylinders that have varying amounts of water in them. The widths of the cylinders and the heights to which they have been filled with water vary. The cylinders all have a hole cut in the side. All of the holes are the same size and they are all at the same height above the bases of the cylinders. There are corks in all of the holes.

Rank these situations, from greatest to smallest, on the basis of the pressure on the cork by the water.

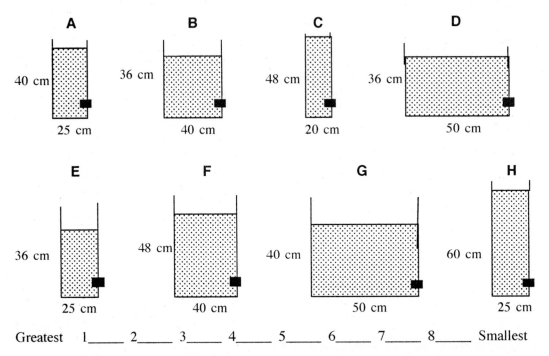

Greatest   1_____ 2_____ 3_____ 4_____ 5_____ 6_____ 7_____ 8_____ Smallest

Or, the pressure is the same for all these cases. _____

Please carefully explain your reasoning.

How sure were you of your ranking? (circle one)

Basically guessed   Sure   Very Sure

1   2   3   4   5   6   7   8   9   10

Cylinders of Liquids—Pressure on the Plug II [108]

The figures below show side views of six hollow cylinders that have varying amounts of various liquids in them. The widths of the cylinders and the heights to which they have been filled with liquid vary. The densities of the liquids, the widths of the cylinders, and the heights to which the cylinders have been filled are given in each figure. The cylinders all have a hole cut in the side. All of the holes are the same size and they are all at the same height above the bases of the cylinders. There are corks in all of the holes.

Rank these situations from greatest to smallest on the basis of the pressure on the cork by the water.

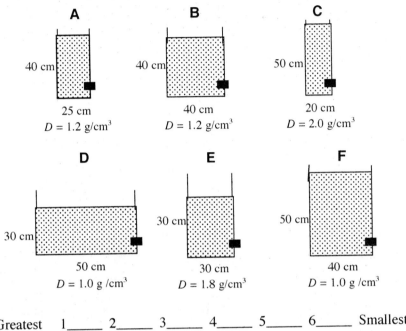

A	B	C
40 cm	40 cm	50 cm
25 cm	40 cm	20 cm
$D = 1.2$ g/cm^3	$D = 1.2$ g/cm^3	$D = 2.0$ g/cm^3

D	E	F
30 cm	30 cm	50 cm
50 cm	30 cm	40 cm
$D = 1.0$ g /cm^3	$D = 1.8$ g/cm^3	$D = 1.0$ g /cm^3

Greatest   1_____ 2_____ 3_____ 4_____ 5_____ 6_____ Smallest

Or, the pressure on the cork by the water is the same for each case. _____

Please carefully explain your reasoning.

How sure were you of your ranking? (circle one)

Basically guessed				Sure				Very Sure	
1	2	3	4	5	6	7	8	9	10

Heat and Thermodynamics Ranking Tasks

You have six styrofoam cups containing the same amount of water at 20°C. You also have six copper blocks whose masses and initial temperatures vary as shown below. One block goes into each cup. (Assume the mass of the water is between 500 g and 1000g.)

Rank these cups according to the maximum temperature of the water after the block is added.

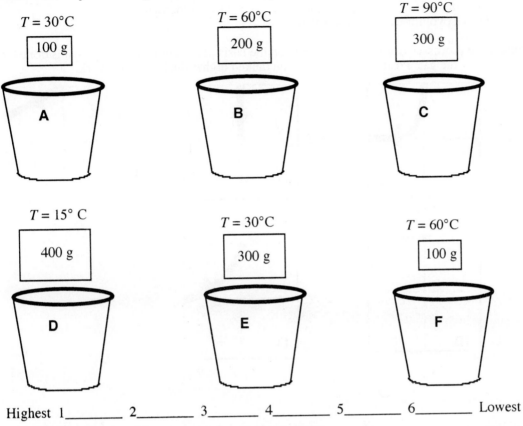

Highest 1_____ 2_____ 3_____ 4_____ 5_____ 6_____ Lowest

Or, all of the cups have the same maximum temperature. _____

Please carefully explain your reasoning.

How sure were you of your ranking? (circle one)
Basically Guessed   Sure   Very Sure
1   2   3   4   5   6   7   8   9   10

Various Thermodynamic Processes—Work Done by Gas [110]

Five thermodynamic processes are illustrated below. All five processes are for the same ideal gas starting at the same initial pressure and temperature (P_o and T_o). Each of the five distinctly different processes results in different final equilibrium states. The labels on each of the diagrams are to be interpreted as follows:

A = adiabatic   L = linear   IT = isothermal
IV = isovolumetric   IB = isobaric

Rank these processes from greatest to least on the basis of the amount of work that is done by the gas. Positive work should be ranked higher than negative work.

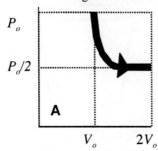

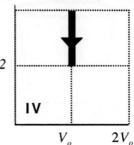

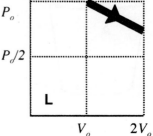

Greatest Work 1______ 2______ 3______ 4______ 5______ Least Work

Or, all of the diagrammed processes require the same amount of work done by the gas. ____

Or, there is no work done by the gas in any of these processes. ____

Please carefully explain your reasoning.

How sure were you of your ranking? (circle one)

Basically Guessed   Sure   Very Sure
1 2 3 4 5 6 7 8 9 10

[110] J. Popp

Gas cylinders with the same mass (or number of molecules) of He gas have various volumes and temperatures, which are given in the diagram below. All of the containers are made of the same material.

Rank these containers from the highest to lowest on the basis of their gauge pressures.

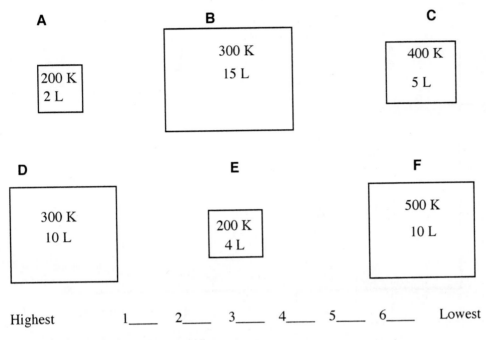

Highest   1____   2____   3____   4____   5____   6____   Lowest

Or, all of the containers have the same guage pressure. _____

Please carefully explain your reasoning.

How sure were you of your ranking? (circle one)

Basically Guessed				Sure				Very Sure	
1	2	3	4	5	6	7	8	9	10

Rank the temperatures of one mole of an ideal gas at the different points identified on the *P-V* diagram below.

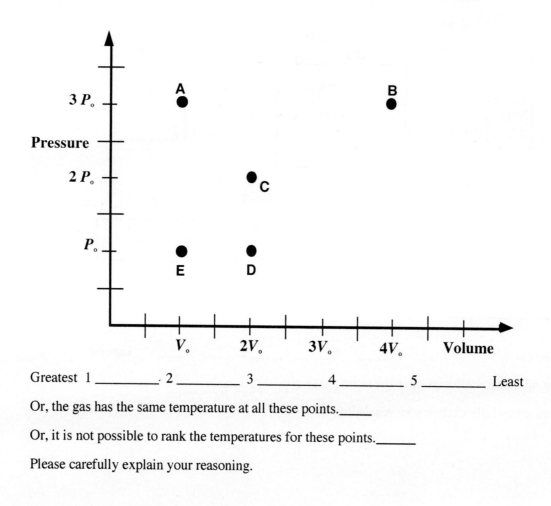

Greatest 1 _____ 2 _____ 3 _____ 4 _____ 5 _____ Least

Or, the gas has the same temperature at all these points._____

Or, it is not possible to rank the temperatures for these points._____

Please carefully explain your reasoning.

How sure were you of your ranking? (circle one)

Basically Guessed				Sure				Very Sure	
1	2	3	4	5	6	7	8	9	10

Rank the temperatures of one mole of the an ideal gas at the different points identified on the
P-V diagram below.

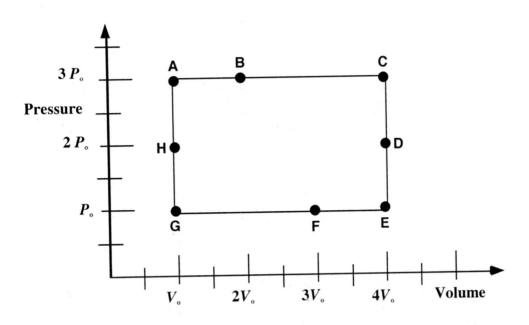

Greatest 1 _____ 2 _____ 3 _____ 4 _____ 5 _____ 6 _____ 7 _____ 8 _____ Least

Or, the gas has the same temperature at all these points._____

Or, it is not possible to rank the temperatures for these points._____

Please carefully explain your reasoning.

How sure were you of your ranking? (circle one)
Basically Guessed   Sure   Very Sure
1   2   3   4   5   6   7   8   9   10

Rank the total internal energies of the ideal gases below that contain a different number (N) of molecules at various temperatures (T) and pressures (P). Note: The volumes are not given; they may or may not be the same.

A

$P=$ 2 atm
$T=$ 200 K
$N=$ 10,000

B

$P=$ 2 atm
$T=$ 200 K
$N=$ 5,000

C

$P=$ 2 atm
$T=$ 200 K
$N=$ 20,000

D

$P=$ 2 atm
$T=$ 300 K
$N=$ 10,000

E

$P=$ 2 atm
$T=$ 150 K
$N=$ 10,000

F

$P=$ 1 atm
$T=$ 150 K
$N=$ 10,000

G

$P=$ 1 atm
$T=$ 300 K
$N=$ 10,000

H

$P=$ 3 atm
$T=$ 200 K
$N=$ 20,000

Greatest   1____   2____   3____   4____   5____   6____   7____   8____   Least

Or, all these gases have the same internal energy. _____

Or, it is not possible to rank the internal energies for these gases. _____

Please carefully explain your reasoning.

How sure were you of your ranking? (circle one)
Basically guessed   Sure   Very Sure
1   2   3   4   5   6   7   8   9   10

Pressure, Volume, and Molecules—Temperature [115]

Rank the temperatures of the ideal gases below that contain a different number of molecules (N) at various pressures (P) and volumes (V).

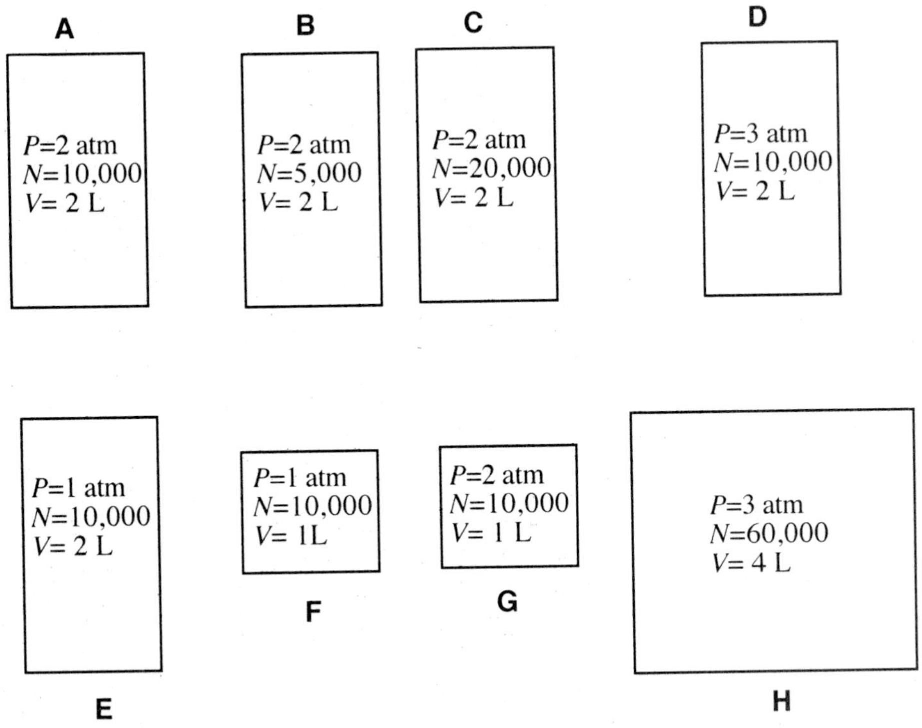

A
P=2 atm
N=10,000
V= 2 L
B
P=2 atm
N=5,000
V= 2 L
C
P=2 atm
N=20,000
V= 2 L
D
P=3 atm
N=10,000
V= 2 L
P=1 atm
N=10,000
V= 2 L
E
P=1 atm
N=10,000
V= 1L
F
P=2 atm
N=10,000
V= 1 L
G
P=3 atm
N=60,000
V= 4 L
H

Greatest 1____ 2____ 3____ 4____ 5____ 6____ 7____ 8____ Least

Or, all these gases have the same temperature. _____

Or, it is not possible to rank the temperatures for these gases. ____________

Please carefully explain your reasoning.

How sure were you of your ranking? (circle one)

Basically guessed				Sure					Very Sure
1	2	3	4	5	6	7	8	9	10

[115] C. Hieggelke

Internal Energy, Volume, and Molecules—Temperature [116]

Rank the temperatures of the ideal gases below that contain various amounts of internal energy (U) and various numbers of molecules (N) in various volumes.

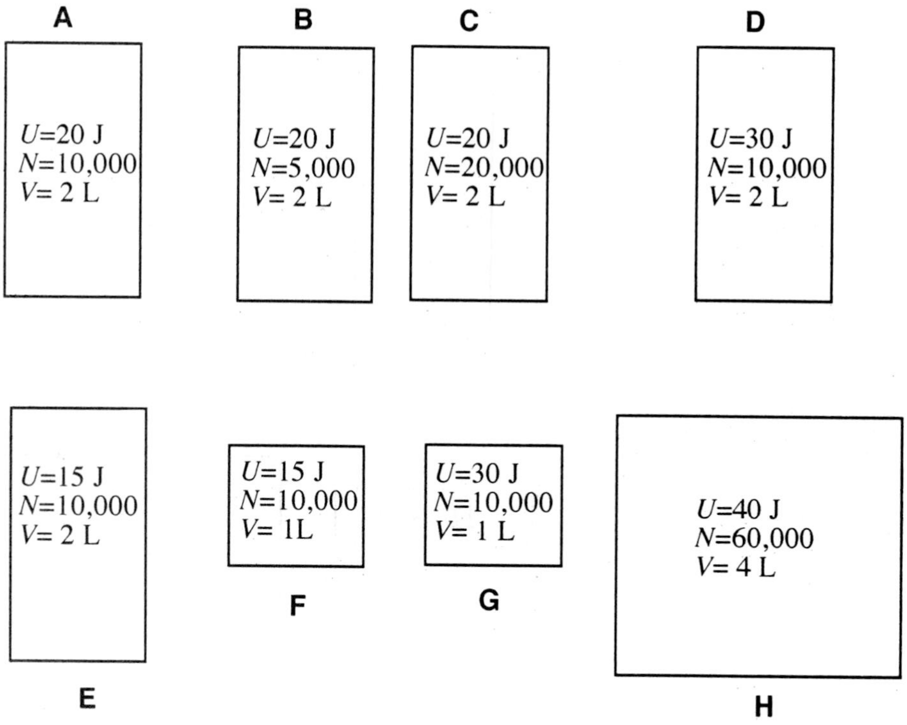

A
U=20 J
N=10,000
V= 2 L
B
U=20 J
N=5,000
V= 2 L
C
U=20 J
N=20,000
V= 2 L
D
U=30 J
N=10,000
V= 2 L
U=15 J
N=10,000
V= 2 L
E
U=15 J
N=10,000
V= 1L
F
U=30 J
N=10,000
V= 1 L
G
U=40 J
N=60,000
V= 4 L
H

Greatest 1____ 2____ 3____ 4____ 5____ 6____ 7____ 8____ Least

Or, all these gases have the same temperature._____

Or, it is not possible to rank the temperatures for these gases.____________

Please carefully explain your reasoning.

How sure were you of your ranking? (circle one)

Basically guessed — Sure — Very Sure

1 2 3 4 5 6 7 8 9 10

---

[116] C. Hieggelke

Internal Energy, Volume, and Molecules—Pressure [117]

Rank the pressures of the ideal gases below that contain various amounts of internal energy (U) and various numbers of molecules (N) in various volumes (V).

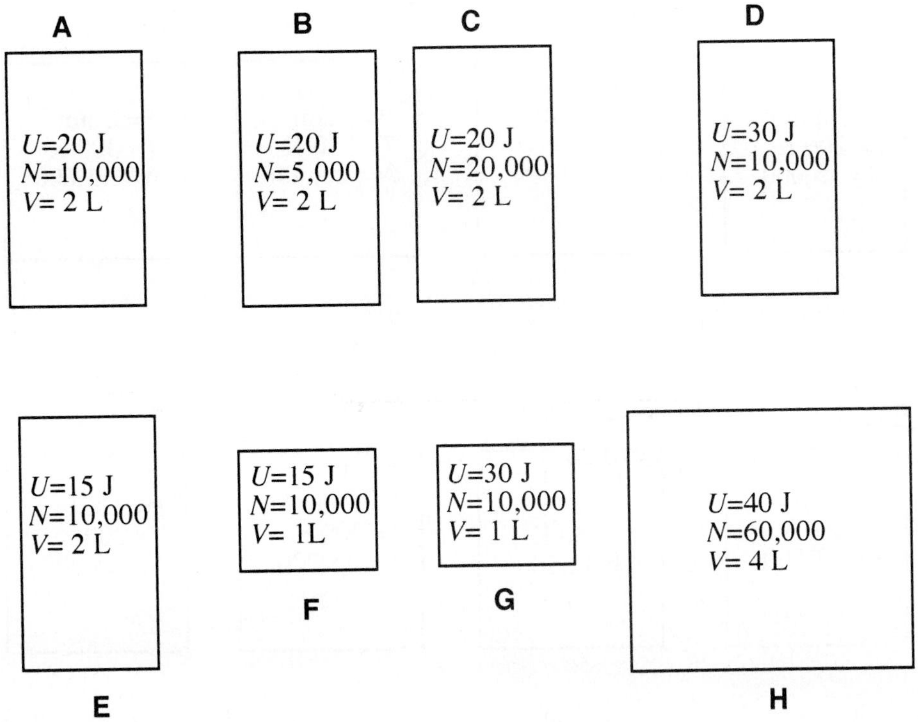

A
U=20 J
N=10,000
V= 2 L
B
U=20 J
N=5,000
V= 2 L
C
U=20 J
N=20,000
V= 2 L
D
U=30 J
N=10,000
V= 2 L
U=15 J
N=10,000
V= 2 L
E
U=15 J
N=10,000
V= 1L
F
U=30 J
N=10,000
V= 1 L
G
U=40 J
N=60,000
V= 4 L
H

Greatest 1___ 2___ 3___ 4___ 5___ 6___ 7___ 8___ Least

Or, all these gases have the same pressure. ______

Or, it is not possible to rank the pressures for these gases. ______

Please carefully explain your reasoning.

How sure were you of your ranking? (circle one)

Basically guessed				Sure					Very Sure
1	2	3	4	5	6	7	8	9	10

[117] C. Hieggelke

Rank the total internal energies of the ideal gases below that contain different number of molecules (N) at various temperatures (T) and pressures (P). Note: The volumes are not given.

A

> $P = 2$ atm
> $T = 200$ K
> $N = 15,000$

B

> $P = 2$ atm
> $T = 150$ K
> $N = 10,000$

C

> $P = 2$ atm
> $T = 250$ K
> $N = 20,000$

D

> $P = 2$ atm
> $T = 300$ K
> $N = 15,000$

E

> $P = 2$ atm
> $T = 150$ K
> $N = 12,500$

F

> $P = 1$ atm
> $T = 150$ K
> $N = 12,500$

G

> $P = 1$ atm
> $T = 300$ K
> $N = 15,000$

H

> $P = 3$ atm
> $T = 200$ K
> $N = 25,000$

Greatest   1____   2____   3____   4____   5____   6____   7____   8____   Least

Or, all these gases the same total internal energy. _____

Or, it is not possible to rank the total internal energies for these gases. _____

Please carefully explain your reasoning.

How sure were you of your ranking? (circle one)

Basically guessed				Sure				Very Sure	
1	2	3	4	5	6	7	8	9	10

Wave Ranking Tasks

Shown below are six waves, which are all the same kind of wave (e.g., all seismic waves) traveling in various media. The waves all have the same frequency, but their amplitudes, A, and wavelengths, λ, vary as shown in the figures. Specific values for these properties are given in each figure.

Rank these waves from greatest to least based on the speed of the waves. That is, put first the wave that is moving fastest and put last the wave that is moving slowest.

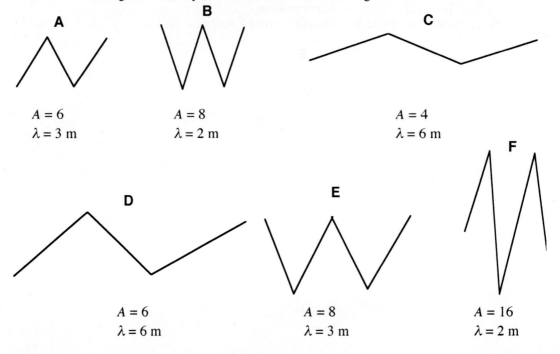

A

$A = 6$
$\lambda = 3$ m

B

$A = 8$
$\lambda = 2$ m

C

$A = 4$
$\lambda = 6$ m

D

$A = 6$
$\lambda = 6$ m

E

$A = 8$
$\lambda = 3$ m

F

$A = 16$
$\lambda = 2$ m

Fastest 1_____ 2_____ 3_____ 4_____ 5_____ 6_____ Slowest

Or, all of these waves travel at the same speed. ____

Or, all of these waves are at rest. ____

Please carefully explain your reasoning.

How sure were you of your ranking? (circle one)
Basically Guessed   Sure   Very Sure
1   2   3   4   5   6   7   8   9   10

Shown below are six wave pulses of either triangular or square shape. These pulses, which vary in amplitude, are all sent down identical ropes under equal tension. The ropes are all the same length, and there is no distortion of the pulses as they travel down the ropes.

Rank these pulses, from greatest to least, on the basis of how long it takes the leading edge to travel 3 m. That is, put first the pulse that takes the most time for the leading edge to travel 3 m, and put last the pulse that travels 3 m in the shortest time.

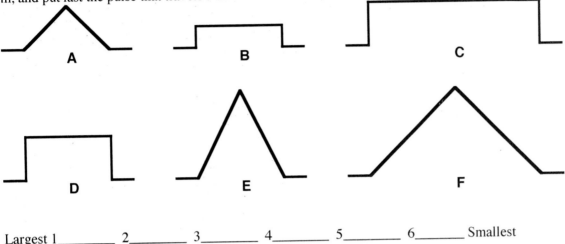

Largest 1_____ 2_____ 3_____ 4_____ 5_____ 6_____ Smallest

Or, all six of these pulses take the same time to travel 3 m. _____

Please carefully explain your reasoning.

How sure were you of your ranking? (circle one)

Basically Guessed   Sure   Very Sure

1   2   3   4   5   6   7   8   9   10

The figures below show systems of standing waves set up in strings, fixed at both ends, under tension. All of the strings are identical except for their lengths and are under the same tension. The variables in these situations, in addition to the lengths (L) of the strings, are the amplitudes (A) at the antinodes and the number of nodes.

Rank these systems, from greatest to least, on the basis of the frequencies of the waves.

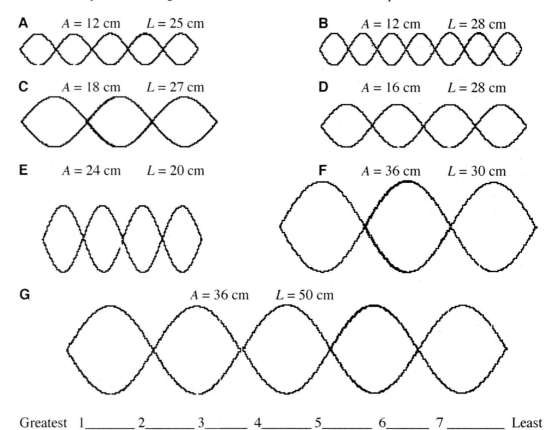

A $A = 12$ cm $L = 25$ cm

B $A = 12$ cm $L = 28$ cm

C $A = 18$ cm $L = 27$ cm

D $A = 16$ cm $L = 28$ cm

E $A = 24$ cm $L = 20$ cm

F $A = 36$ cm $L = 30$ cm

G $A = 36$ cm $L = 50$ cm

Greatest 1_____ 2_____ 3_____ 4_____ 5_____ 6_____ 7 _____ Least

Or, all of these systems have the same frequency. _____

Please carefully explain your reasoning.

How sure were you of your ranking? (circle one)

Basically Guessed					Sure				Very Sure
1	2	3	4	5	6	7	8	9	10

[121] C. Hieggelke, D. Maloney, T. O'Kuma

The figures below show systems of standing waves set up in strings, fixed at both ends, under tension. All of the strings are identical except for their lengths and are under the same tension. The variables in these situations, in addition to the lengths (L) of the strings, are the amplitudes (A) at the antinodes and the number of nodes.

Rank these systems, from greatest to least, on the basis of the wavelengths of the waves.

A $A = 12$ cm $L = 25$ cm

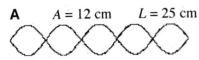

B $A = 12$ cm $L = 28$ cm

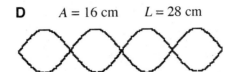

C $A = 18$ cm $L = 27$ cm

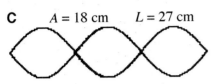

D $A = 16$ cm $L = 28$ cm

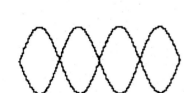

E $A = 24$ cm $L = 20$ cm

F $A = 36$ cm $L = 30$ cm

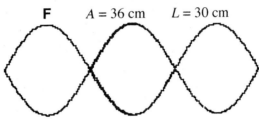

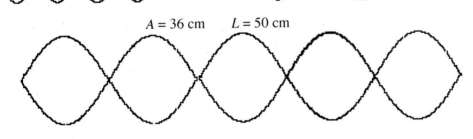

G $A = 36$ cm $L = 50$ cm

Greatest 1_____ 2_____ 3_____ 4_____ 5_____ 6_____ 7_____ Least

Or, all of these systems have the same wavelength. _____

Please carefully explain your reasoning.

How sure were you of your ranking? (circle one)
Basically Guessed   Sure   Very Sure
1   2   3   4   5   6   7   8   9   10

Shown below are six standing wave systems in strings. These systems vary in frequency of oscillation, tension in the strings, and number of nodes. The systems are also set up in various strings. The specific values for the string tensions and the frequencies of oscillation are given in each figure. All of the strings have the same length.

Rank these systems, from greatest to least, on the basis of the speeds of the waves in the strings. That is, put first the system whose waves have the greatest speed in their string and put last the system whose waves are traveling slowest in their string.

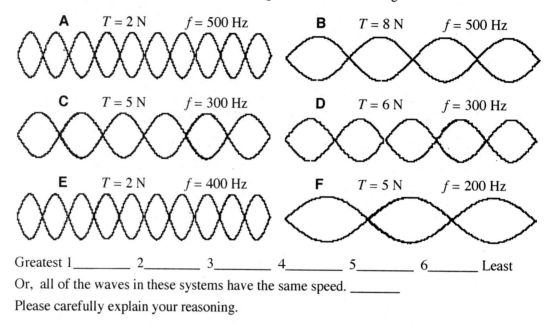

A	$T = 2\,N$	$f = 500\,Hz$	B	$T = 8\,N$	$f = 500\,Hz$
C	$T = 5\,N$	$f = 300\,Hz$	D	$T = 6\,N$	$f = 300\,Hz$
E	$T = 2\,N$	$f = 400\,Hz$	F	$T = 5\,N$	$f = 200\,Hz$

Greatest 1_____ 2_____ 3_____ 4_____ 5_____ 6_____ Least

Or, all of the waves in these systems have the same speed. _____

Please carefully explain your reasoning.

How sure were you of your ranking? (circle one)

Basically Guessed   Sure   Very Sure

1   2   3   4   5   6   7   8   9   10

[123] C. Hieggelke, D. Maloney, T. O'Kuma

Shown below are six waves of equal wavelengths traveling in the same medium. Rank these waves from greatest to lowest according to their speed in the medium. Assume the waves are all the same type.

A

B

C

D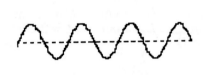

E

F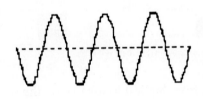

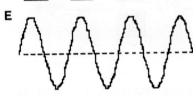

Greatest 1_____ 2_____ 3_____ 4_____ 5_____ 6_____ Lowest

Or, all of the waves have the same wave speed. _____

Please carefully explain your reasoning.

How sure were you of your ranking? (circle one)
Basically Guessed   Sure   Very Sure
1   2   3   4   5   6   7   8   9   10

Pairs of Transverse Waves—Superposition[125]

Shown below are six different pairs of rectangular transverse wave pulses that vary in height (*H*) and length (*L*). Specific values for the heights and lengths are given in each figure for each pulse. In each pair the pulses are moving toward each other. At some point in time the pulses meet and interact (interfere) with each other.

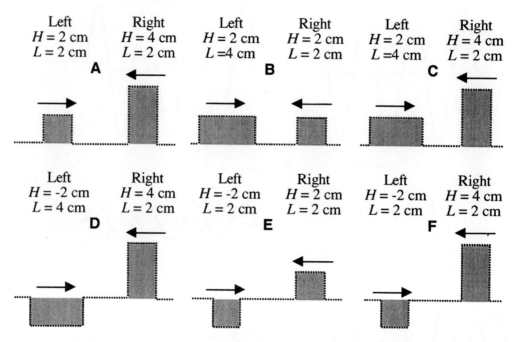

Rank these pairs, from highest to lowest, on the basis of the maximum (peak) height of the combined (resultant) pulse at the center of the combined waves at the time when the centers of the pairs coincide.

Greatest   1 _____ 2 _____ 3 _____ 4 _____ 5 _____ 6 _____ Least

Or, all of these pairs will have the same maximum height._____

Please carefully explain your reasoning.

How sure were you of your ranking? (circle one)

Guessed					Sure				Very Sure
1	2	3	4	5	6	7	8	9	10

[125] C. Hieggelke, D. Maloney, T. O'Kuma

Shown are six waves of equal wavelengths traveling in the same medium. Rank these waves from highest to lowest according to their energy in the medium. Assume the waves are all the same type.

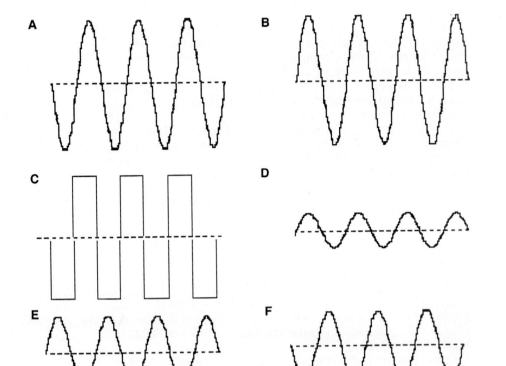

Greatest 1_____ 2_____ 3_____ 4_____ 5_____ 6_____ Least

Or, all waves have the same wave average energy. _____

Please carefully explain your reasoning.

How sure were you of your ranking? (circle one)

Basically Guessed					Sure			Very Sure	
1	2	3	4	5	6	7	8	9	10

Electrostatics Ranking Tasks

Given below are seven arrangements of two electric charges. In each figure, a point labeled P is also identified. All of the charges are the same size, 20 C, but they can be either positive or negative. The charges and point P all lie on a straight line. The distances between adjacent items, either between two charges or between a charge and point P, are all 5 cm. There are no other charges in this region. For this problem, we are going to place a +5 C charge at point P.

Rank these arrangements from greatest to least on the basis of the strength of the electric force on the +5 C charge when it is placed at point P. That is, put first the arrangement that will exert the strongest force on the +5 C charge at point P, and put last the arrangement that will exert the weakest force on the +5 C charge when it is placed at point P.

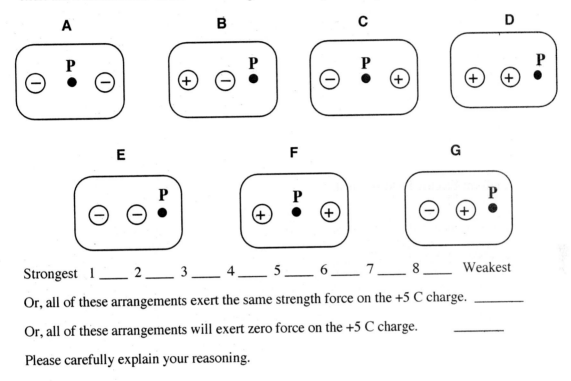

Strongest 1 ____ 2 ____ 3 ____ 4 ____ 5 ____ 6 ____ 7 ____ 8 ____ Weakest

Or, all of these arrangements exert the same strength force on the +5 C charge. _____

Or, all of these arrangements will exert zero force on the +5 C charge. _____

Please carefully explain your reasoning.

How sure were you of your ranking? (circle one)

Basically Guessed				Sure				Very Sure	
1	2	3	4	5	6	7	8	9	10

Three Linear Electric Charges—Electric Force[127]

Given below are arrangements of three fixed electric charges. In each figure, a point labeled P is also identified. All of the charges are the same size charge, q, but they can be either positive or negative as indicated. The charges and point P all lie on a straight line. The distances between adjacent items, either between two charges or between a charge and point P, are all the same. There are no other charges in this region. A test charge, $+Q$, is placed at point P.

Rank these arrangements from greatest to least on the basis of the strength (magnitude) of the electric force on the test charge, $+Q$, at P.

A ⊕   ⊕   ⊕   P   **B** ⊕   ⊕   P   ⊕

C ⊕   ⊖   P   ⊕   **D** ⊕   ⊕   P   ⊖

E ⊕   ⊕   ⊖   P   **F** ⊕   ⊖   ⊕   P

Greatest 1 _____ 2 _____ 3 _____ 4 _____ 5 _____ 6 _____ Least

Or, all of these arrangements exert the same magnitude force on the $+Q$ test charge. _____

Or, all of these arrangements will exert zero force on the $+Q$ test charge. _____

Please carefully explain your reasoning.

How sure were you of your ranking? (circle one)

Basically Guessed   Sure   Very Sure
1   2   3   4   5   6   7   8   9   10

Two Nonlinear Electric Charges—Electric Force[128]

Given below are arrangements of two fixed electric charges. In each figure, a point labeled P is also identified. All of the charges are the same size, q, but they can be either positive or negative as indicated. The distances between adjacent items, either between two charges or between a charge and point P, are all the same. There are no other charges in this region. For this problem, we are going to place a test charge, $+Q$, at point P.

Rank these arrangements from greatest to least on the basis of the strength (magnitude) of the electric force on the test charge, $+Q$, at P.

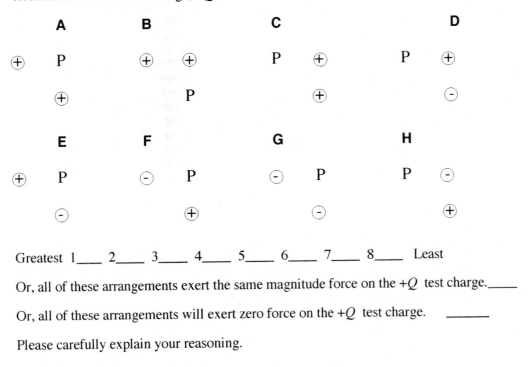

Greatest 1____ 2____ 3____ 4____ 5____ 6____ 7____ 8____ Least

Or, all of these arrangements exert the same magnitude force on the $+Q$ test charge.____

Or, all of these arrangements will exert zero force on the $+Q$ test charge.   _____

Please carefully explain your reasoning.

How sure were you of your ranking? (circle one)

Basically Guessed   Sure   Very Sure

1   2   3   4   5   6   7   8   9   10

Shown below are eight hollow spheres of different sizes made of an electrically conducting material such as copper. On each sphere there is a charge, as given in the figure, which is distributed evenly over the sphere. Each figure is independent of the others (they do not affect each other).

Rank these situations, from greatest to least, on the basis of the magnitude of the electric field at the center of the sphere.

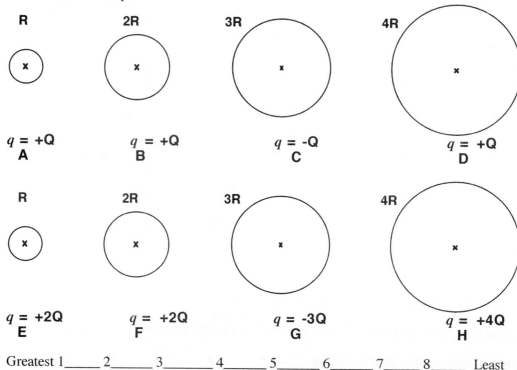

Greatest 1_____ 2_____ 3_____ 4_____ 5_____ 6_____ 7_____ 8_____ Least

Or, the magnitude of the electric field at the center is the same (but not zero) for these cases.___

Or, the magnitude of the electric field at the center is zero for these cases. ___

Please carefully explain your reasoning.

How sure were you of your ranking? (circle one)

Basically Guessed   Sure   Very Sure

1   2   3   4   5   6   7   8   9   10

Charged Conducting Spheres—Electric Potential at the Center [130]

Shown below are eight hollow spheres of different sizes made of an electrically conducting material such as copper. On each sphere there is a charge, as given in the figure, which is distributed evenly over the sphere. Each figure is independent of the others (they do not affect each other).

Rank these situations, from greatest to least, on the basis of the electric potential at the center of the sphere.

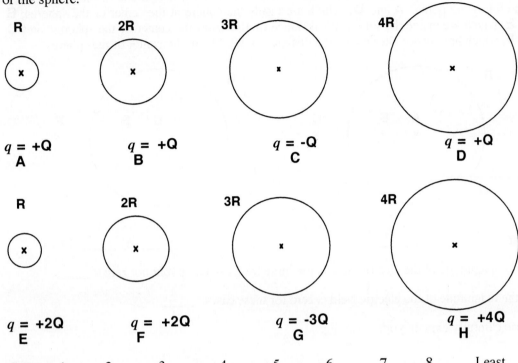

$q = +Q$
A

$q = +Q$
B

$q = -Q$
C

$q = +Q$
D

$q = +2Q$
E

$q = +2Q$
F

$q = -3Q$
G

$q = +4Q$
H

Greatest 1_____ 2_____ 3_____ 4_____ 5_____ 6_____ 7_____ 8_____ Least

Or, the electric potential at the center is the same (but not zero) for these cases. _____

Or, the electric potential at the center is zero for these cases. _____

Please carefully explain your reasoning.

How sure were you of your ranking? (circle one)
Basically Guessed   Sure   Very Sure
1   2   3   4   5   6   7   8   9   10

[130] C. Hieggelke, T. O'Kuma
Ranking Task Exercises in Physics

Electricity and Magnetism

Shown below are two hollow spheres made of an electrically conducting material such as copper. On these spheres there is a different charge on each as given in the figure which is distributed evenly over the sphere. Each figure or sphere is independent of the others (they do not affect each other).

Rank these situations, from greatest to least, on the basis of the magnitude of the electric field at the following points: **A** and **D,** which are inside the sphere at the center of the spheres; **B** and **E,** which are inside the sphere at a distance of R/2 from the center of the spheres; and **C** and **F,** which are outside the sphere at a distance of 3R/2 from the center of the spheres.

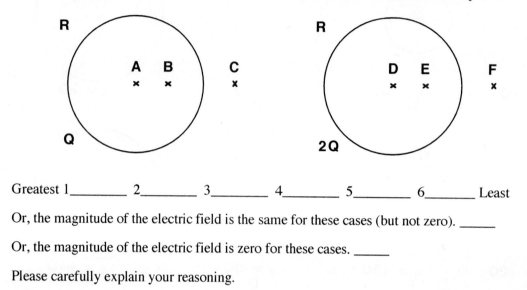

Greatest 1_____ 2_____ 3_____ 4_____ 5_____ 6_____ Least

Or, the magnitude of the electric field is the same for these cases (but not zero). _____

Or, the magnitude of the electric field is zero for these cases. _____

Please carefully explain your reasoning.

How sure were you of your ranking? (circle one)
Basically Guessed   Sure   Very Sure
1   2   3   4   5   6   7   8   9   10

Charged Conducting Spheres—Electric Potential at Various Points[132]

Shown below are two hollow spheres made of an electrically conducting material such as copper. On these spheres there is a different charge on each, as given in the figure, which is distributed evenly over the sphere. Each figure or sphere is independent of the others (they do not affect each other).

Rank these situations, from greatest to least, on the basis of the electric potential at the following points: **A** and **D,** which are inside the sphere at the center of the spheres; **B** and **E,** which are inside the sphere at a distance of R/2 from the center of the spheres; and **C** and **F,** which are outside the sphere at a distance of 3R/2 from the center of the spheres.

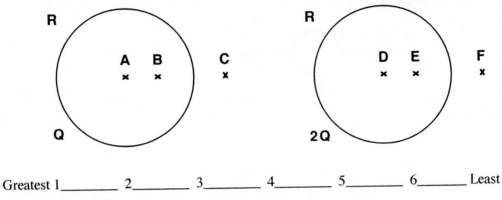

Greatest 1_____ 2_____ 3_____ 4_____ 5_____ 6_____ Least

Or, the electric potential is the same for these cases (but not zero). _____

Or, the electric potential is zero for these cases. _____

Please carefully explain your reasoning.

How sure were you of your ranking? (circle one)

Basically Guessed				Sure				Very Sure	
1	2	3	4	5	6	7	8	9	10

[132] C. Hieggelke, T. O'Kuma
Ranking Task Exercises in Physics

Shown below are six hollow spheres made of an electrically conducting material such as copper. At the center of each sphere is a charge q, which is the same sign and magnitude for all six cases. Outside these spheres, at various distances, are electric charges of various magnitudes. Given in each figure is the magnitude of the outside charges, as well as the distance between that charge and the sphere. Each figure is independent of the others (they do not affect each other).

Rank these situations, from greatest to least, on the basis of the magnitude of the force on the charge at the center of the sphere. That is, put first the situation where the charge at the center of the sphere experiences the strongest force, and put last the situation where the charge in the center experiences the weakest force.

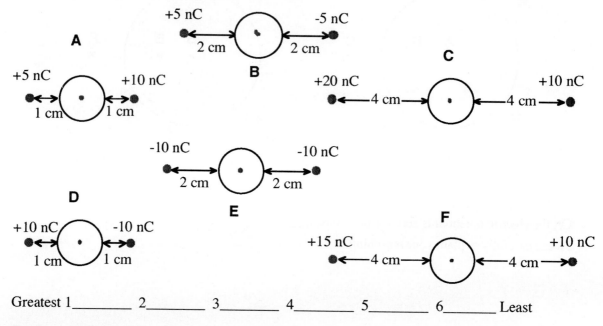

Greatest 1_____ 2_____ 3_____ 4_____ 5_____ 6_____ Least

Or, the magnitude of the electric force on the center charge is the same for these cases. _____

Or, the magnitude of the electric force on the center charge is zero for these cases. _____

Please carefully explain your reasoning.

How sure were you of your ranking? (circle one)
Basically Guessed   Sure   Very Sure
1   2   3   4   5   6   7   8   9   10

Shown below are six identical hollow spheres with a radius of 1 cm made of an electrically conducting material such as copper. Outside these spheres, at various distances, are two electric charges of various magnitudes. Some of the charges are positive and some are negative. Given in each figure is the sign and magnitude of the charge, as well as the distance of the charge from the sphere. Each figure is independent of the others (they do not affect each other).

Rank these situations, from greatest to least, on the basis of the strengths (magnitudes) of the electric fields at the centers of the spheres.

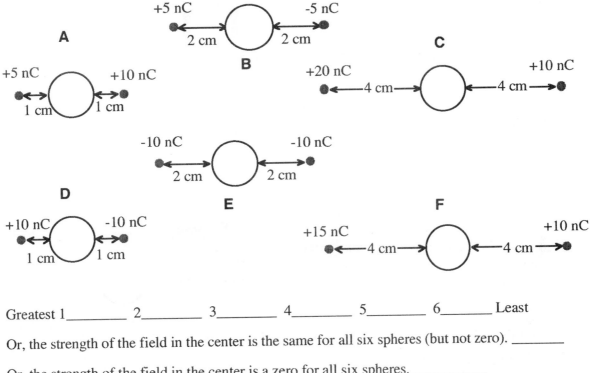

Greatest 1_____ 2_____ 3_____ 4_____ 5_____ 6_____ Least

Or, the strength of the field in the center is the same for all six spheres (but not zero). _____

Or, the strength of the field in the center is a zero for all six spheres. _____

Please carefully explain your reasoning.

How sure were you of your ranking? (circle one)
Basically Guessed   Sure   Very Sure
1   2   3   4   5   6   7   8   9   10

The following diagrams show three separate situations involving charges on metal spheres that are initially in contact. The positively charged rod is brought up to the same distance from each set of metal spheres. The spheres are then separated simultaneously from each other using the insulating handles. Finally the charged rod is removed. (Assume the separation is carried out in vacuum.)

Rank the charge on each sphere from most positive to most negative after the rod is removed.

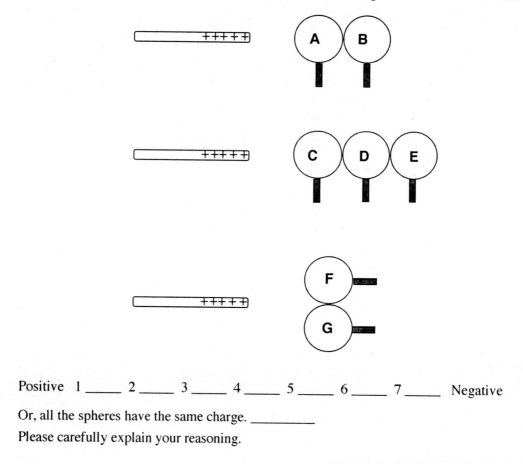

Positive 1 _____ 2 _____ 3 _____ 4 _____ 5 _____ 6 _____ 7 _____ Negative

Or, all the spheres have the same charge. _____

Please carefully explain your reasoning.

How sure were you of your ranking? (circle one)
Basically Guessed   Sure   Very Sure
1   2   3   4   5   6   7   8   9   10

A (positively or negatively) charged rod is brought up to the same distance from each set of metal spheres as shown in separate situations below. The spheres in each pair are initially in contact, but they are then separated while the rod is still in place. Then the rod is removed.

Rank the net charge on each sphere from most positive to most negative after the spheres have been separated and the charged rod removed.

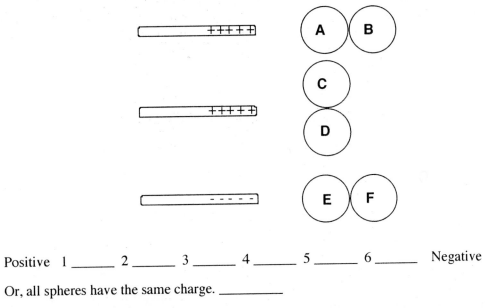

Positive 1 _____ 2 _____ 3 _____ 4 _____ 5 _____ 6 _____ Negative

Or, all spheres have the same charge. _____

Please carefully explain your reasoning.

How sure were you of your ranking? (circle one)
Basically Guessed   Sure   Very Sure
1   2   3   4   5   6   7   8   9   10

Pairs of Charges—Attractive and Repulsive Force [137]

The following diagrams show three separate pairs of point charges. The pairs only interact with each other since they are far apart.

Rank the force on each point charge from most attractive to most repulsive.

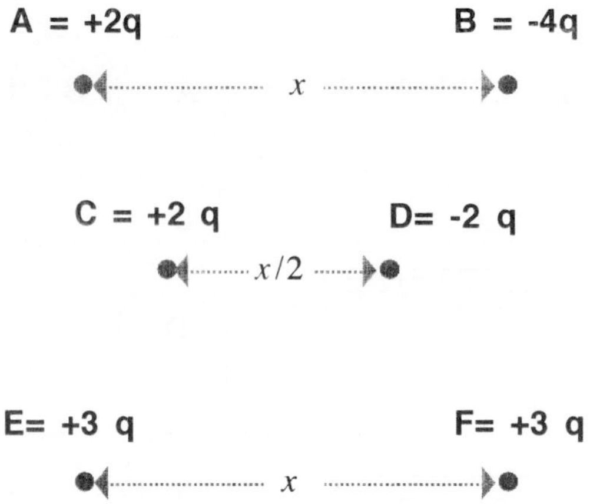
A = +2q
B = -4q
x
C = +2 q
D= -2 q
x/2
E= +3 q
F= +3 q
x

Highest 1________ 2________ 3________ 4________ 5________ 6________ Lowest

Or, all of these charges experience the same force. _________

Please carefully explain your reasoning.

How sure were you of your ranking? (circle one)

Basically Guessed					Sure				Very Sure
1	2	3	4	5	6	7	8	9	10

[137] B. Emerson

Suspended Charges—Angle [138]

The figure below shows a charged sphere that is suspended from a string in a uniform electric field that is pointing in the horizontal direction. Six possible combinations of sphere mass and electric charge are listed in the chart below. All of these spheres are suspended in the same uniform field and the gravitational field is directed downward. When the spheres are placed in the field they will swing away from vertical and hang at an angle θ from the vertical as shown in the diagram.

Rank, from greatest to least, the angle θ the string will form with the vertical for these different spheres.

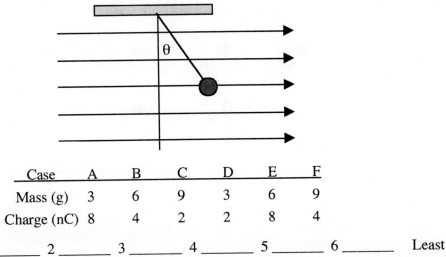

Case	A	B	C	D	E	F
Mass (g)	3	6	9	3	6	9
Charge (nC)	8	4	2	2	8	4

Greatest   1 _____ 2 _____ 3 _____ 4 _____ 5 _____ 6 _____ Least

Or, all of the angles will be the same. _____

Please carefully explain your reasoning.

How sure were you of your ranking? (circle one)
Basically Guessed   Sure   Very Sure
1   2   3   4   5   6   7   8   9   10

[138] G. Fazzari, A. Weiner, T. Bennie-George, O. Selgrad, C. Hieggelke

We have a large region of space that has a uniform electric field in the $+x$ direction ($\Rightarrow$). At the point $(0,0)$ m, the electric field is 30 **i** N/C and the electric potential is 100 volts.

Rank from greatest to least the strength (magnitude) of the electric force on a $+5$ C charge when it is placed at rest at each of the following points.

A: (0, 6) m   **B**: (0, 3) m   **C**: (-3, 6) m   **D**: (3, 6) m   **E**: (3,3) m   **F**: (6, 6) m

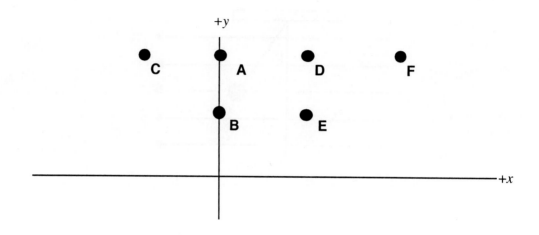

Greatest   1 _____   2 _____   3 _____   4 _____   5 _____   6 _____   Least

Or, the 5 C charge will experience the same strength electric force at all of these points. _____

Or, the 5 C charge will not experience a force at any of these points. _____

Please carefully explain your reasoning.

How sure were you of your ranking? (circle one)

Basically Guessed   Sure   Very Sure

1   2   3   4   5   6   7   8   9   10

Uniform Electric Field—Electric Force on Charge at Rest II [140]

We have a large region of space that has a uniform electric field in the +x direction (⇒) as indicated by the arrows in the diagram below. At the point (0,0) *m*, the electric field is 30 **i** N/C and the electric potential is 100 volts.

Rank from greatest to least the strength (magnitude) of the electric force on a +5 C charge when it is placed at rest at each of the following points.

A: (0, 6) *m*   **B**: (0, 3) *m*   **C**: (-3, 6) *m*   **D**: (3, 6) *m*   **E**: (3,3) *m*   **F**: (6, 6) *m*

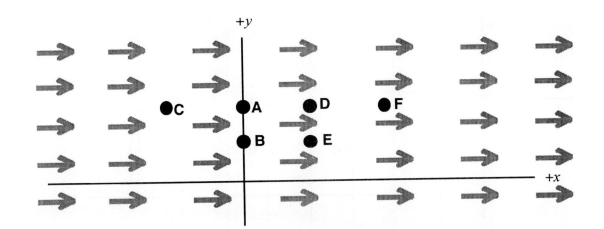

Greatest   1 ____   2 ____   3 ____   4 ____   5 ____   6 ____   Least

Or, the 5 C charge will experience the same strength electric force at all of these points. _____

Or, the 5 C charge will not experience a force at any of these points. _____

Please carefully explain your reasoning.

How sure were you of your ranking? (circle one)
Basically Guessed   Sure   Very Sure
1   2   3   4   5   6   7   8   9   10

We have a large region of space that has a uniform electric field in the +*x* direction (⇒). In the diagram below we show the equipotential lines for this field. At the point (0,0) m, the electric field is 30 **i** N/C and the electric potential is 100 volts.

Rank from greatest to least the strength (magnitude) of the electric force on a +5 C charge when it is placed at rest at each of the following points.

A: (0, 6) m   **B**: (0, 3) m   **C**: (-3, 6) m   **D**: (3, 6) m   **E**: (3,3) m   **F**: (6, 6) m

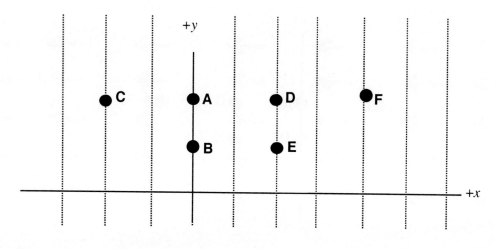

Greatest   1 ____   2 ____   3 ____   4 ____   5 ____   6 ____   Least

Or, the 5 C charge will experience the same strength electric force at all of these points. _____

Or, the 5 C charge will not experience a force at any of these points. _____

Please carefully explain your reasoning.

How sure were you of your ranking? (circle one)

Basically Guessed   Sure   Very Sure

1   2   3   4   5   6   7   8   9   10

Uniform Electric Field—Electric Force on Various Charges at Rest [142]

We have a large region of space that has a uniform electric field in the $+x$ direction (⇒) as indicated by the arrows in the diagram below. At the point (0,0) m, the electric field is 30 **i** N/C and the electric potential is 100 volts.

Rank from greatest to least the strength (magnitude) of the electric force on the charges listed when placed at rest at the points specified below. Each charge is placed at its specified point separately.

A: (0, 6) m	B: (0, 3) m	C: (-3, 6) m	D: (3, 6) m	E: (3,3) m	F: (6, 6) m
$q = +2$ C	$q = +2$ C	$q = +2$ C	$q = +4$ C	$q = +4$ C	$q = +2$ C

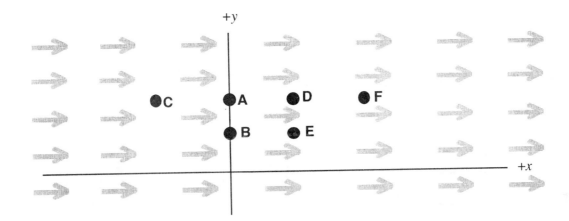

Greatest 1_____ 2_____ 3_____ 4_____ 5_____ 6_____ Least

Or, all of the charges have the same strength electric force acting on them._______

Or, none of these charges would experience a force. _________

Please carefully explain your reasoning.

How sure were you of your ranking? (circle one)

Basically Guessed				Sure				Very Sure	
1	2	3	4	5	6	7	8	9	10

---

[142] C. Hieggelke

Given below are seven arrangements of two electric charges. In each figure, a point labeled P is also identified. All of the charges are the same size, Q, but they can be either positive or negative. The charges and point P all lie on a straight line. The distances between adjacent items, either between two charges or between a charge and point P, are all x cm. There is no charge at point P, nor are there any other charges in this region.

Rank these arrangements from greatest to least on the basis of the strength of the electric field at point P. That is, put first the arrangement that produces the strongest field at point P, and put last the arrangement that produces the weakest field at point P.

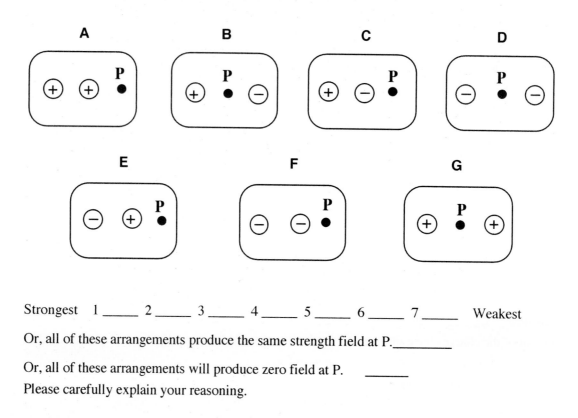

Strongest   1 _____ 2 _____ 3 _____ 4 _____ 5 _____ 6 _____ 7 _____   Weakest

Or, all of these arrangements produce the same strength field at P._____

Or, all of these arrangements will produce zero field at P.   _____

Please carefully explain your reasoning.

How sure were you of your ranking? (circle one)
Basically Guessed   Sure   Very Sure
1   2   3   4   5   6   7   8   9   10

Electron Within a Charged Capacitor—Force on the Electron [144]

Consider an electron between the plates of a charged capacitor. The figures below show situations where the potential across the capacitor, and the separations between the capacitor plates, vary. Specific values are given in each figure.

Rank according to the magnitude of the force felt by the electron.

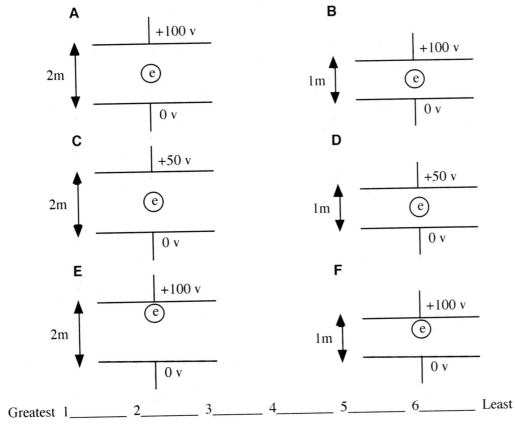

Greatest 1_____ 2_____ 3_____ 4_____ 5_____ 6_____ Least

Or, all of the forces are the same strength. _____

Please carefully explain your reasoning.

How sure were you of your ranking? (circle one)

Basically Guessed   Sure   Very Sure

1   2   3   4   5   6   7   8   9   10

[144] A. Van Heuvelen, S. Heath, B. Willis, L. Bryant
Ranking Task Exercises in Physics   155   Electricity and Magnetism

In each of the following situations two conducting spheres with the same size are shown with an initial given number of units of charge. The two spheres are brought into contact with each other. After several moments the spheres are separated.

Rank the situations as to the quantity of charge on the first (left) sphere from the highest positive charge to the lowest negative charge after they have been separated. (Note that -6 is lower than -2.)

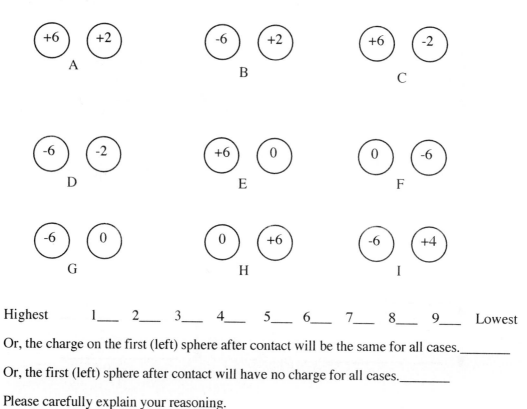

Highest   1___ 2___ 3___ 4___ 5___ 6___ 7___ 8___ 9___   Lowest

Or, the charge on the first (left) sphere after contact will be the same for all cases._____

Or, the first (left) sphere after contact will have no charge for all cases._____

Please carefully explain your reasoning.

How sure were you of your ranking? (Circle one)
Basically Guessed   Sure   Very Sure
1   2   3   4   5   6   7   8   9   10

145 J. Gundlach, B. Kaasa, U. Pandey, M. West
Ranking Task Exercises in Physics   156   Electricity and Magnetism

In each of the following situations two conducting spheres with the same size are shown with an initial given number of units of charge. The two spheres are brought into contact with each other. After several moments the spheres are separated.

Rank the situations as to the quantity of charge on the second (right) sphere from the highest positive charge to the lowest negative charge after they have been separated. (Note that -6 is lower than -2.)

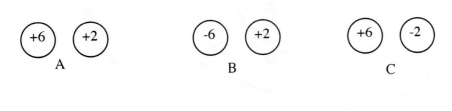

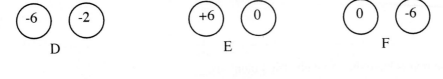

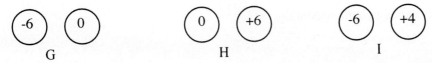

Highest   1___ 2___ 3___ 4___ 5___ 6___ 7___ 8___ 9___ Lowest

Or, the charge on the second (right) sphere after contact will be the same for all cases._____

Or, the second (right) sphere after contact will have no charge for all cases.____

Please carefully explain your reasoning.

How sure were you of your ranking? (Circle one)
Basically Guessed   Sure   Very Sure
1   2   3   4   5   6   7   8   9   10

A non-uniform electric field is being represented below by electric field lines. Six points in this region are identified in this diagram.

Rank the electric potential of the marked points from greatest to least.

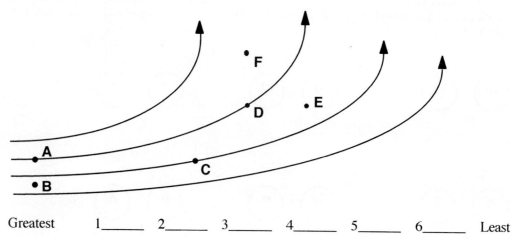

Greatest   1_____ 2_____ 3_____ 4_____ 5_____ 6_____ Least

Or, all of these points have the same electric potentials._____

Please carefully explain your reasoning.

How sure were you of your ranking? (Circle one)

Guessed				Sure				Very Sure	
1	2	3	4	5	6	7	8	9	10

A non-uniform electric field is being represented below by electric field lines. Six points in this region are identified in this diagram.

Rank the strength (magnitude) of the electric field of the marked points from greatest to least.

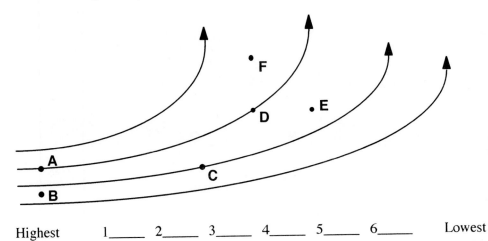

Highest   1_____   2_____   3_____   4_____   5_____   6_____   Lowest

Or, the magnitude of the electric field is the same at all of these points._____

Please carefully explain your reasoning.

How sure were you of your ranking? (Circle one)
Basically Guessed   Sure   Very Sure
1   2   3   4   5   6   7   8   9   10

Two parallel plates that have been charged create a uniform electric field of 30 N/C between the plates. Rank the electrical potential differences of all the different combinations listed below between the four points **M** at (2,0) m; **N** at (5, 0)m; **O** at (8,0) m; and **P** at (2, 3) m within this region. (Positive values are larger than negative values.)

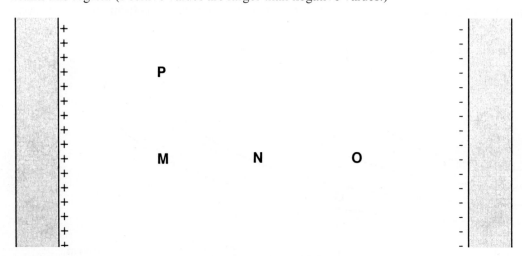

Potential difference combinations that are to be ranked—

A: M to **N**	**B: M** to **O**	**C: N** to **O**	**D: P** to **M**
E: P to **N**	**F: P** to **O**	**G: N** to **M**	**H: M** to **P**

Highest   1____ 2____ 3____ 4____ 5____ 6____ 7____ 8____ Lowest

Or, all the combinations have the same potential difference._____

Please carefully explain your reasoning.

How sure were you of your ranking? (Circle one of the following.)
Basically Guessed   Sure   Very Sure
1   2   3   4   5   6   7   8   9   10

Uniform Electric Field—Strength of the Electric Field I [150]

We have a large region of space that has a uniform electric field in the +x direction (⇒). At the point (0,0) m, the electric field is 30 **i** N/C and the electric potential is 100 volts.

Rank the strength (magnitude) of the electric field from greatest to least at the following points within this region.

A: (0, 6) m   **B**: (0, 3) m   **C**: (-3, 6) m   **D**: (3, 6) m   **E**: (3,3) m   **F**: (6, 6) m

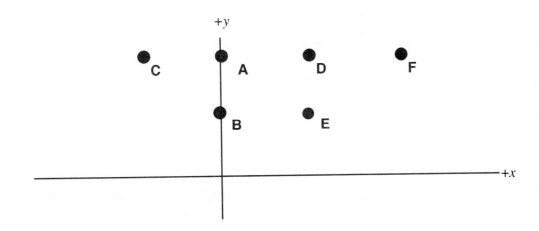

Greatest   1_____   2_____   3_____   4_____   5_____   6_____   Least

Or, all of the points have the same electric field magnitude._____

Please carefully explain your reasoning.

How sure were you of your ranking? (circle one)

Basically Guessed   Sure   Very Sure
1   2   3   4   5   6   7   8   9   10

Uniform Electric Field—Strength of the Electric Field II[151]

We have a large region of space that has a uniform electric field in the $+x$ direction (⇒) as indicated by the arrows in the diagram below. At the point (0,0) m, the electric field is 30 **i** N/C and the electric potential is 100 volts.

Rank the strength (magnitude) of the electric field from greatest to least at the following points within this region.

A: (0, 6) m **B**: (0, 3) m **C**: (-3, 6) m **D**: (3, 6) m **E**: (3,3) m **F**: (6, 6) m

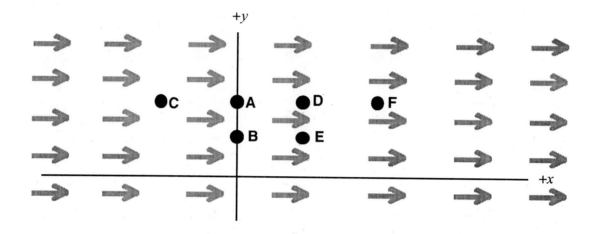

Greatest 1_____ 2_____ 3_____ 4_____ 5_____ 6_____ Least

Or, all of the points have the same electric field magnitude._______

Please carefully explain your reasoning.

How sure were you of your ranking? (circle one)

Basically Guessed Sure Very Sure

1 2 3 4 5 6 7 8 9 10

[151] C. Hieggelke

Uniform Electric Field/Potential Lines—Strength of the Electric Field [152]

We have a large region of space that has a uniform electric field in the $+x$ direction (⇒). In the diagram below we show the equipotential lines for this field. At the point (0,0) m, the electric field is 30 **i** N/C and the electric potential is 100 volts.

Rank the strength (magnitude) of the electric field from greatest to least at the following points within this region.

A: (0, 6) m **B**: (0, 3) m **C**: (-3, 6) m **D**: (3, 6) m **E**: (3,3) m **F**: (6, 6) m

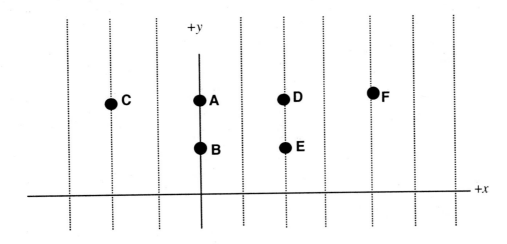

Greatest 1_____ 2_____ 3_____ 4_____ 5_____ 6_____ Least

Or, all of the points have the same electric field magnitude.________

Please carefully explain your reasoning.

How sure were you of your ranking? (circle one)

Basically Guessed				Sure				Very Sure	
1	2	3	4	5	6	7	8	9	10

[152] C. Hieggelke

We have a large region of space that has a uniform electric field in the +x direction (⇒) as indicated by the arrows in the diagram below. At the point (0,0) *m*, the electric field is 30 **i** N/C and the electric potential is 100 volts.

Rank the electric potential from greatest to least at the following points within this region.

A: (0, 6) *m*   **B**: (0, 3) *m*   **C**: (-3, 6) *m*   **D**: (3, 6) *m*   **E**: (3,3) *m*   **F**: (6, 6) *m*

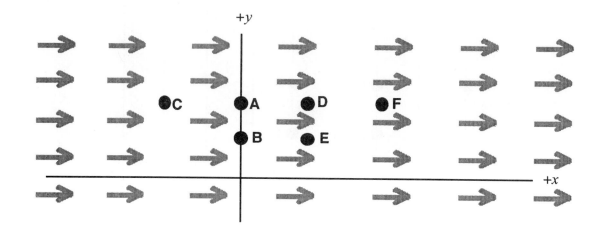

Greatest   1_____   2_____   3_____   4_____   5_____   6_____   Least

Or, all the points have the same electric potential. _____

Please carefully explain your reasoning.

How sure were you of your ranking? (circle one)
Basically Guessed   Sure   Very Sure
1   2   3   4   5   6   7   8   9   10

[153] C. Hieggelke
Ranking Task Exercises in Physics   164   Electricity and Magnetism

Uniform Electric Field—Potential Energy of a Positive Charge[154]

We have a large region of space that has a uniform electric field in the +*x* direction (⇒). At the point (0,0) m, the electric field is 30 **i** N/C and the electric potential is 100 volts.

Rank the points specified below on the basis of the electric potential energy of a single charge of +5 C that may be placed at these points.

A: (0, 6) m   **B**: (0, 3) m   **C**: (-3, 6) m   **D**: (3, 6) m   **E**: (3,3) m   **F**: (6, 6) m

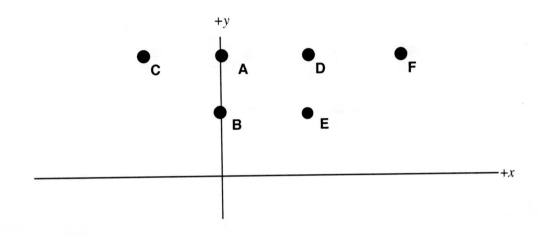

Greatest   1_____   2_____   3_____   4_____   5_____   6_____   Least

Or, the +5C charge would have the same electric potential energy at all of these points. _____

Please carefully explain your reasoning.

How sure were you of your ranking? (circle one)

Basically Guessed					Sure				Very Sure
1	2	3	4	5	6	7	8	9	10

[154] C. Hieggelke
Ranking Task Exercises in Physics

Electricity and Magnetism

Uniform Electric Field—Potential Energy of a Negative Charge[155]

We have a large region of space that has a uniform electric field in the +*x* direction (⇒). At the point (0,0) m, the electric field is 30 **i** N/C and the electric potential is 100 volts.

Rank the points specified below on the basis of the electric potential energy of a single negative charge of -5 C that may be placed at these points.

A: (0, 6) m   **B**: (0, 3) m   **C**: (-3, 6) m   **D**: (3, 6) m   **E**: (3,3) m   **F**: (6, 6) m

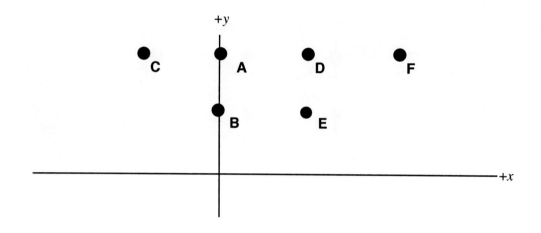

Greatest   1_____   2_____   3_____   4_____   5_____   6_____   Least

Or, the –5 C charge would have the same electric potential energy at all of these points. _____

Please carefully explain your reasoning.

How sure were you of your ranking? (circle one)
Basically Guessed   Sure   Very Sure
1   2   3   4   5   6   7   8   9   10

Uniform Electric Field—Change in Potential Energy of a Positive Charge[156]

We have a large region of space that has a uniform electric field in the +x direction (⇒). At the point (0,0) m, the electric field is 30 **i** N/C and the electric potential is 100 volts.

Rank the points specified below on the basis of the change in the electric potential energy of a single positive charge of +5 C that is moved from the origin (0,0) to these particular points. That is, put first the point that will involve the largest change in electric potential energy as the charge is moved from (0,0) to that point, and put last the point that will involve the smallest change in electric potential energy as the charge is moved from (0,0) to that point. Note that the some of these changes may be negative and that -5 < 2.

A: (0, 6) m   **B**: (0, 3) m   **C**: (-3, 6) m   **D**: (3, 6) m   **E**: (3,3) m   **F**: (6, 6) m

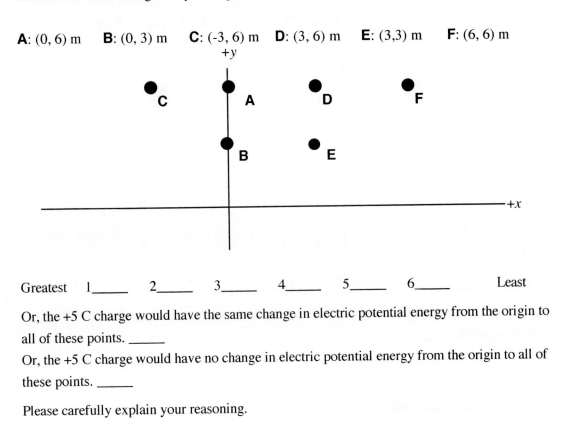

Greatest   1_____   2_____   3_____   4_____   5_____   6_____   Least

Or, the +5 C charge would have the same change in electric potential energy from the origin to all of these points. _____

Or, the +5 C charge would have no change in electric potential energy from the origin to all of these points. _____

Please carefully explain your reasoning.

Uniform Electric Field—Change in Potential Energy of a Negative Charge[157]

We have a large region of space that has a uniform electric field in the +x direction (⇒). At the point (0,0) m, the electric field is 30 **i** N/C and the electric potential is 100 volts.

Rank the points specified below on the basis of the change in the electric potential energy of a single negative charge of -5 C that is moved from the origin (0,0) to these particular points. That is, put first the point that will involve the largest change in electric potential energy as the charge is moved from (0,0) to that point, and put last the point that will involve the smallest change in electric potential energy as the charge is moved from (0,0) to that point. Note that the some of these changes may be negative and that -5 < 2.

A: (0, 6) m   **B**: (0, 3) m   **C**: (-3, 6) m   **D**: (3, 6) m   **E**: (3,3) m   **F**: (6, 6) m

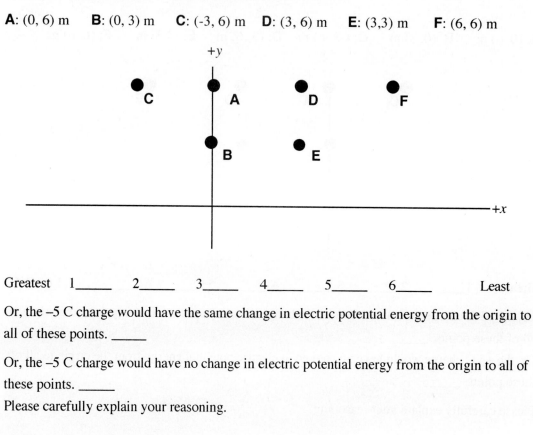

Greatest   1_____   2_____   3_____   4_____   5_____   6_____   Least

Or, the –5 C charge would have the same change in electric potential energy from the origin to all of these points. _____

Or, the –5 C charge would have no change in electric potential energy from the origin to all of these points. _____

Please carefully explain your reasoning.

How sure were you of your ranking? (circle one)
Basically Guessed   Sure   Very Sure
1   2   3   4   5   6   7   8   9   10

Electron Within a Charged Capacitor—Electric Potential Energy [158]

Consider an electron between the plates of a charged capacitor. The figures below show situations where the potential across the capacitor, and the separations between the capacitor plates, vary. Specific values are given in each figure.

Rank according to the magnitude of the electric potential energy of the electron.

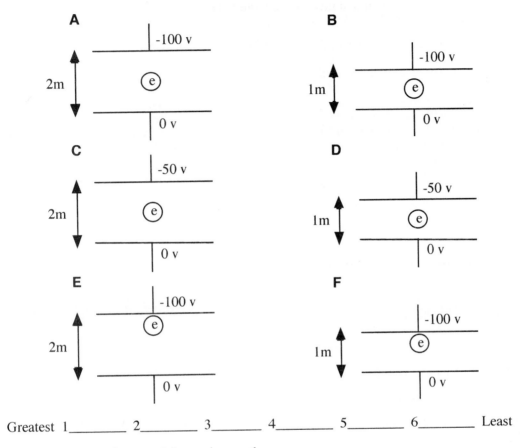

Greatest 1_____ 2_____ 3_____ 4_____ 5_____ 6_____ Least

Or, all of the electric potential energies are the same. _____

Please carefully explain your reasoning.

How sure were you of your ranking? (circle one)

Basically Guessed   Sure   Very Sure

1   2   3   4   5   6   7   8   9   10

[158] A. Van Heuvelen, S. Heath, B. Willis, L. Bryant, D. Maloney, T. O'Kuma

DC Circuit Ranking Tasks

Basic Circuits with Identical Capacitors—Charge on a Capacitor [159]

Shown below are eight capacitor circuits. All of the capacitors are identical, and all are fully charged. The batteries are also identical. In each circuit, one capacitor is labeled X.

Rank these circuits in terms of the charge on capacitor X. That is, put first the circuit in which capacitor X has the largest charge, and put last the circuit in which capacitor X has the smallest charge. If two or more circuits result in identical charges on capacitor X, give these circuits equal ranking.

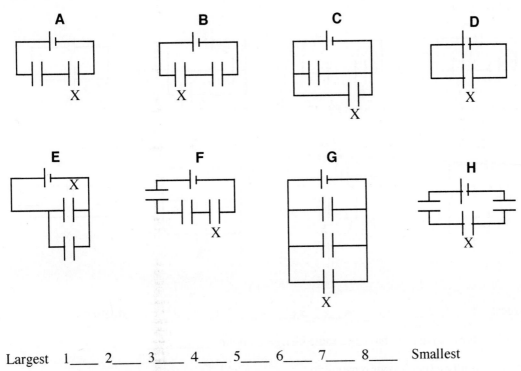

Largest   1____ 2____ 3____ 4____ 5____ 6____ 7____ 8____   Smallest

Or, all of these capacitors have the same charge on them. _____

Please carefully explain your reasoning.

How sure were you of your ranking? (circle one)

Basically Guessed					Sure				Very Sure	
1	2	3	4	5	6	7	8	9	10

[159] L. Takahashi, D. Maloney, C. Hieggelke
Ranking Task Exercises in Physics

Electricity and Magnetism

Circuits with Identical Capacitors—Charge on a Capacitor

Shown below are eight capacitor circuits. All of the capacitors are identical, and all are fully charged. The batteries are also identical. In each circuit, one capacitor is labeled X.

Rank these circuits in terms of the charge on capacitor X. That is, put first the circuit in which capacitor X has the largest charge, and put last the circuit in which capacitor X has the smallest charge. If two or more circuits result in identical charges on capacitor X, give these circuits equal ranking.

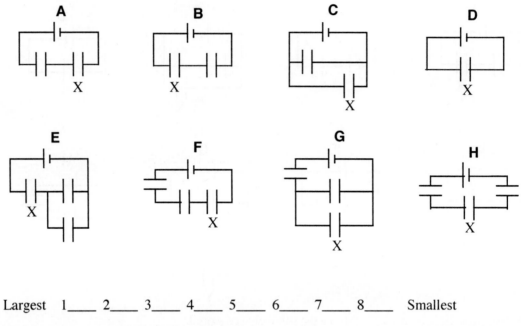

Largest   1____   2____   3____   4____   5____   6____   7____   8____   Smallest

Or, all of these capacitors have the same charge on them. _____

Please carefully explain your reasoning.

How sure were you of your ranking? (circle one)
Basically Guessed   Sure   Very Sure
1   2   3   4   5   6   7   8   9   10

Shown below are eight capacitor circuits. All of the capacitors are either 1 μF or 2 μF, and all are fully charged. The batteries are also identical. In each circuit, one capacitor is labeled X.

Rank these circuits in terms of the charge stored in capacitor X. That is, put first the circuit in which capacitor X has the largest charge stored, and put last the circuit in which capacitor X has the smallest charge stored. If two or more circuits result in identical charge stored for capacitor X, give these circuits equal ranking.

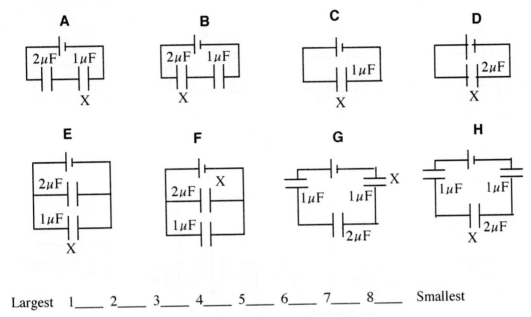

Largest 1____ 2____ 3____ 4____ 5____ 6____ 7____ 8____ Smallest

Or, the capacitors marked X all have the same charge stored. _____

Please carefully explain your reasoning.

How sure were you of your ranking? (circle one)

Basically Guessed				Sure				Very Sure	
1	2	3	4	5	6	7	8	9	10

[161] L. Takahashi, C. Hieggelke
Ranking Task Exercises in Physics

Electricity and Magnetism

Basic Circuits with Identical Capacitors—Voltage Across a Capacitor[162]

Shown below are eight capacitor circuits. All of the capacitors are ,and all are fully charged. The batteries are also identical. In each circuit, one capacitor is labeled X.

Rank these circuits in terms of the voltage across capacitor X. That is, put first the circuit in which capacitor X has the largest voltage, and put last the circuit in which capacitor X has the smallest voltage. If two or more circuits result in identical voltages for capacitor X, give these circuits equal ranking.

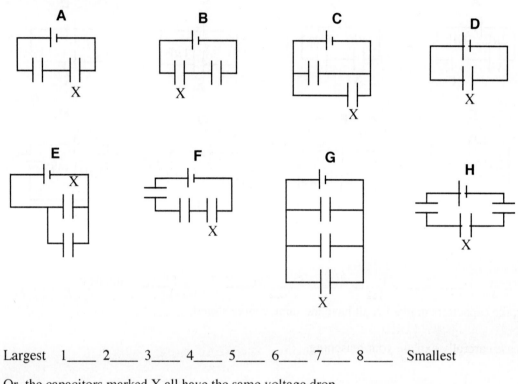

Largest   1____ 2____ 3____ 4____ 5____ 6____ 7____ 8____   Smallest

Or, the capacitors marked X all have the same voltage drop. _____

Please carefully explain your reasoning.

How sure were you of your ranking? (circle one)

Basically Guessed				Sure				Very Sure	
1	2	3	4	5	6	7	8	9	10

[162] L. Takahashi, C. Hieggelke, D. Maloney
Ranking Task Exercises in Physics

Electricity and Magnetism

The circuit shown below has circuit elements α, β, γ, and δ, which are either a 1 F capacitor or a 1, 2, or 4 ohm resistor. Rank the arrangements of circuit elements given below in order from largest to smallest magnitude of current that flows after a long time through the 6 volt battery connected as shown.

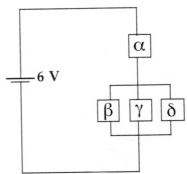

Arrangements	α	β	γ	δ
Arrangement A	1F=1 Farad	4Ω = 4 ohms	1Ω = 1 ohm	2Ω = 2 ohms
Arrangement B	1Ω	1F	2Ω	4Ω
Arrangement C	2Ω	1Ω	4Ω	1F
Arrangement D	4Ω	2Ω	1F	1Ω
Arrangement E	1F	1Ω	2Ω	4Ω
Arrangement F	1Ω	4Ω	2Ω	1F
Arrangement G	2Ω	4Ω	1F	1Ω

Ranking of arrangements by magnitude of current through the battery:

Largest   1_____ 2_____ 3_____ 4_____ 5_____ 6_____ 7_____   Smallest

Which of these arrangements have the same magnitude of current, if any?

_____

Please carefully explain your reasoning.

How sure were you of your ranking? (circle one)
Basically Guessed   Sure   Very Sure
1   2   3   4   5   6   7   8   9   10

The figures below show six segments of wires that are carrying electric currents. In all six cases, the currents are flowing to the right (into the page). As you can see from the values in the figures, the pieces of the wires shown have different lengths, and they are carrying different currents. For the ranking below, we are only interested in the segments of the wires actually shown in the figures.

Rank these segments from greatest to least on the basis of the net electric charge, i.e., the difference in number of positive charges and negative charges in each segment. That is, put first the segment that has the greatest net electric charge on it, and put last the segment that has the smallest net electric charge on it.

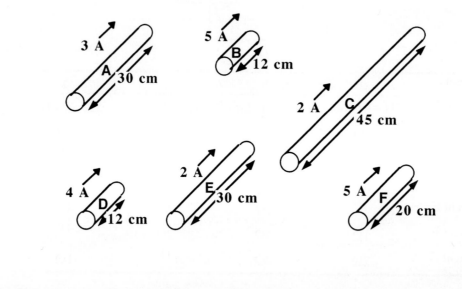

Greatest   1_____ 2_____ 3_____ 4_____ 5_____ 6_____   Smallest

Or, all of these segments have the same net charge. _____

Or, the net charge is zero for all of these segments. _____

Please carefully explain your reasoning.

How sure were you of your ranking? (circle one)
Basically Guessed   Sure   Very Sure
1   2   3   4   5   6   7   8   9   10

Batteries and Bulbs—Bulb Brightness[165]

Identical batteries are connected in different arrangements to the same light bulb. Rank these items on the basis of bulb brightness.

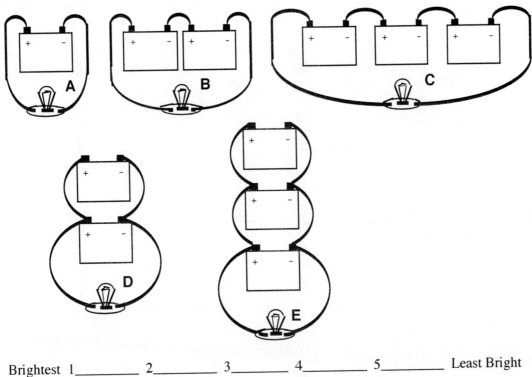

Brightest 1_____ 2_____ 3_____ 4_____ 5_____ Least Bright

Or, all bulbs have the same brightness. _____

Please carefully explain your reasoning.

How sure were you of your ranking? (circle one)
Basically Guessed   Sure   Very Sure
1   2   3   4   5   6   7   8   9   10

All the resistors in the circuits below are identical at 2Ω each. The batteries are ideal with voltages of 6 or 12 volts as shown below. All connecting wires have negligible resistance. Rank the current passing through the upper right hand corner of each circuit from greatest to least.

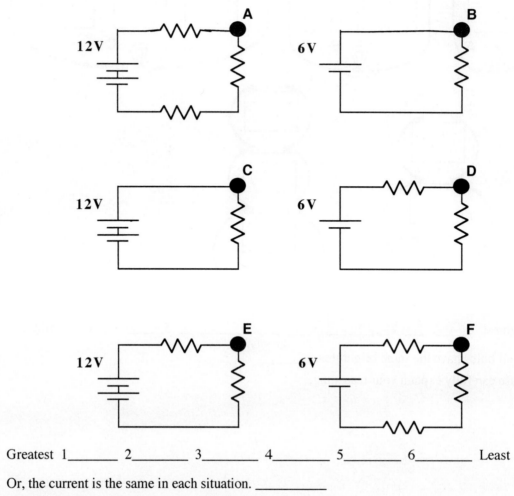

Greatest 1_____ 2_____ 3_____ 4_____ 5_____ 6_____ Least

Or, the current is the same in each situation. _____

Please carefully explain your reasoning.

How sure were you of your ranking? (circle one)
Basically Guessed   Sure   Very Sure
1   2   3   4   5   6   7   8   9   10

Shown below are six segments of wires that are carrying electric currents. All of these segments have the same length and the same diameter, but the wires are made of different materials so they have different resistances. The currents flowing in these segments also vary. Specific values for each of these properties are given in the figures.

Rank these situations from highest to lowest on the basis of the net electric charge on each segment, i.e., on the basis of the difference between the number of positive and negative charges in the wire segment. Put first the situation that has the greatest net charge, and put last the situation with the smallest net charge, i.e., the smallest difference between the number of positive and negative charges.

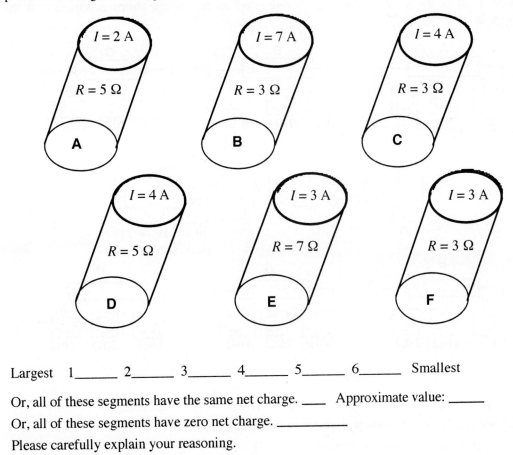

Largest 1_____ 2_____ 3_____ 4_____ 5_____ 6_____ Smallest

Or, all of these segments have the same net charge. ____ Approximate value: _____

Or, all of these segments have zero net charge. _____

Please carefully explain your reasoning.

How sure were you of your ranking? (circle one)

Basically Guessed				Sure				Very Sure	
1	2	3	4	5	6	7	8	9	10

The eight circuits below all have one battery, shown as ▭, and three light bulbs, shown as ▢. All of the batteries are identical. The bulbs can have different resistances (resistance is the opposition a circuit element presents to the flow of current) given in units of ohms, Ω. The specific values for these resistances are given in the figures. One of the three bulbs in each circuit is designated by an X. This is the bulb we are interested in. All of the wires in these circuits are identical and of equal length.

Rank these circuits, from greatest to least, on the basis of the voltage drop across the bulb marked X. That is, put first the circuit where the voltage drop across bulb X is the largest of any of the X bulbs, and put last the circuit where the voltage drop across bulb X is the smallest of all of the X bulbs. We are not interested in the voltage drops across any of the other bulbs.

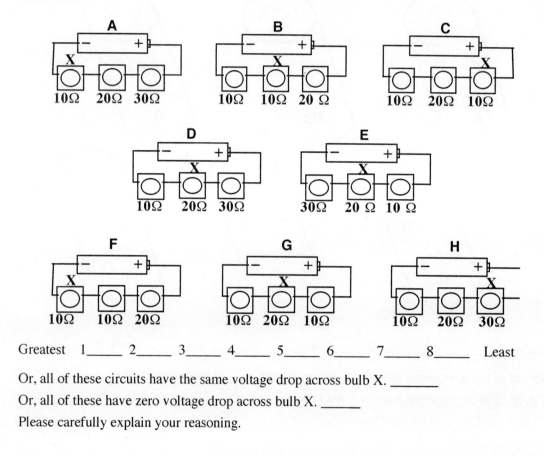

Greatest 1_____ 2_____ 3_____ 4_____ 5_____ 6_____ 7_____ 8_____ Least

Or, all of these circuits have the same voltage drop across bulb X. _____

Or, all of these have zero voltage drop across bulb X. _____

Please carefully explain your reasoning.

How sure were you of your ranking? (circle one)
Basically Guessed   Sure   Very Sure
1   2   3   4   5   6   7   8   9   10

All the resistors in the circuits below are identical at 2Ω each. The batteries are ideal with voltages of 6 or 12 volts as shown. All connecting wires have negligible resistance. Rank the voltage at the upper right-hand corner relative to the ground. Place the highest voltage first and the lowest voltage last.

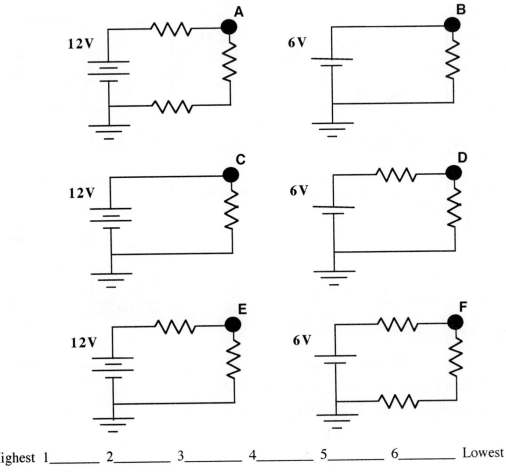

Highest 1_____ 2_____ 3_____ 4_____ 5_____ 6_____ Lowest

Or, all of the voltages are the same. _____

Please carefully explain your reasoning.

How sure were you of your ranking? (circle one)

Basically Guessed					Sure			Very Sure	
1	2	3	4	5	6	7	8	9	10

All resistors shown are identical in the circuits below. The switches in each case are open. Rank the voltmeter readings from highest to lowest.

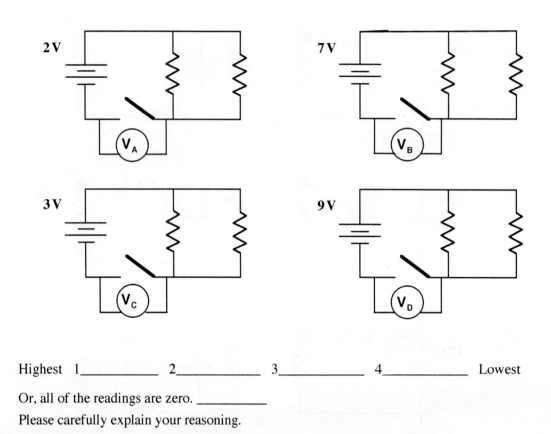

Highest 1_____ 2_____ 3_____ 4_____ Lowest

Or, all of the readings are zero. _____

Please carefully explain your reasoning.

How sure were you of your ranking? (circle one)

Basically Guessed   Sure   Very Sure

1   2   3   4   5   6   7   8   9   10

All resistors shown are identical in the circuits below. The switches in each case are open. Rank the voltmeter readings from highest to lowest.

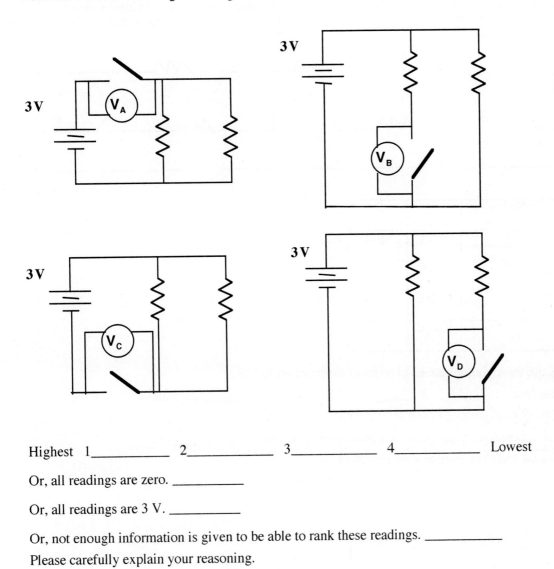

Highest 1_____ 2_____ 3_____ 4_____ Lowest

Or, all readings are zero. _____

Or, all readings are 3 V. _____

Or, not enough information is given to be able to rank these readings. _____

Please carefully explain your reasoning.

How sure were you of your ranking? (circle one)
Basically Guessed   Sure   Very Sure
1   2   3   4   5   6   7   8   9   10

Circuit with Two Open and Closed Switches—Ammeter Readings 172

Shown below is a DC circuit that contains two switches. Each switch is resistanceless when closed. All of the connecting wires should be considered to have zero resistance. All of the resistors shown are identical. The circuit contains an ideal ammeter and an ideal voltmeter. The diagram shows the switches open. Below the diagram are four different switch configurations for the circuit.

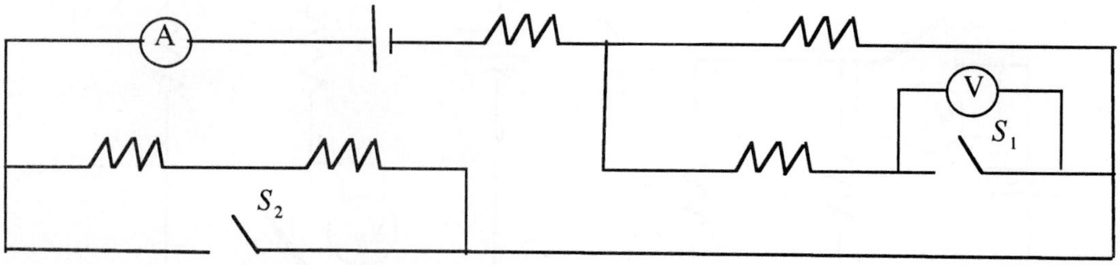

Configuration	S_1	S_2
A	open	open
B	open	closed
C	closed	open
D	closed	closed

Rank these configurations in terms of the ammeter reading.

Largest 1 _____ 2 _____ 3 _____ 4 _____ Smallest

Or, all configurations produce the same ammeter reading. _________

Or, all configurations produce a zero ammeter reading. _________

Please carefully explain your reasoning.

How sure were you of your ranking? (circle one)

Basically Guessed Sure Very Sure

1 2 3 4 5 6 7 8 9 10

---

172 L. Takahashi, C. Hieggelke

Circuit with Two Open and Closed Switches—Voltmeter Readings[173]

Shown below is a DC circuit that contains two switches. Each switch is resistanceless when closed. All of the connecting wires should be considered to have zero resistance. All of the resistors shown are identical. The circuit contains an ideal ammeter and an ideal voltmeter. The diagram shows the switches open. Below the diagram are four different switch configurations for the circuit.

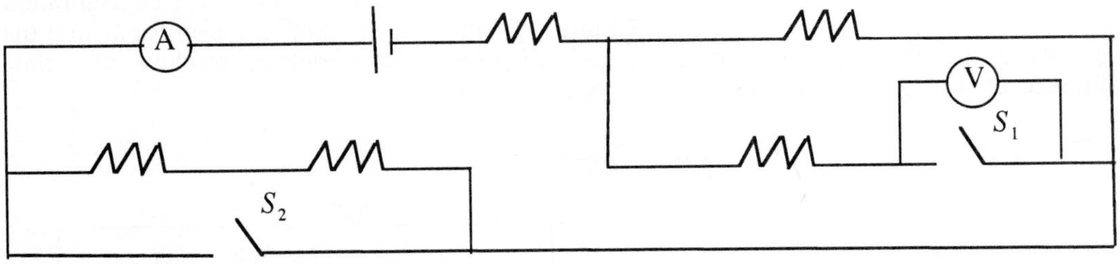

Configuration	S_1	S_2
A	open	open
B	open	closed
C	closed	open
D	closed	closed

Rank these configurations in terms of the voltmeter reading.

Largest 1 _____ 2 _____ 3 _____ 4 _____ Smallest

Or, all configurations produce the same voltmeter reading. _________

Or, all configurations produce a zero voltmeter reading. _________

Please carefully explain your reasoning.

How sure were you of your ranking? (circle one)

Basically Guessed / Sure / Very Sure

1 2 3 4 5 6 7 8 9 10

[173] L. Takahashi, C. Hieggelke
Ranking Task Exercises in Physics

Circuit with Three Open and Closed Switches—Voltmeter Readings I [174]

Shown below is a DC circuit that contains a number of switches. Each switch is resistanceless when closed. All of the connecting wires may be considered to have zero resistance. All of the resistors shown are identical. The circuit contains an ideal voltmeter. The diagram shows all of the switches open. Below the diagram are eight different switch configurations for the circuit.

Rank these configurations in terms of the voltmeter reading. That is, put first the configuration for which the voltmeter gives the largest reading, and put last the configuration for which the voltmeter gives the smallest reading. If two or more configurations produce the same voltmeter reading, give these configurations equal ranking.

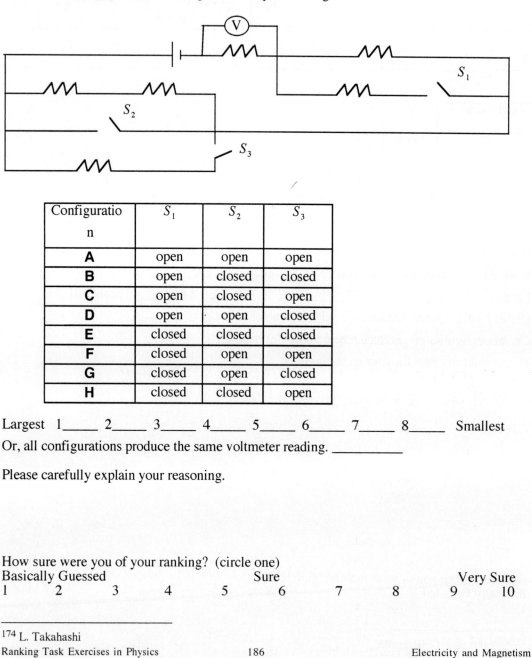

Configuration	S_1	S_2	S_3
A	open	open	open
B	open	closed	closed
C	open	closed	open
D	open	open	closed
E	closed	closed	closed
F	closed	open	open
G	closed	open	closed
H	closed	closed	open

Largest 1_____ 2_____ 3_____ 4_____ 5_____ 6_____ 7_____ 8_____ Smallest

Or, all configurations produce the same voltmeter reading. _____

Please carefully explain your reasoning.

How sure were you of your ranking? (circle one)

Basically Guessed   Sure   Very Sure

1   2   3   4   5   6   7   8   9   10

[174] L. Takahashi

Shown below is a DC circuit that contains a number of switches. Each switch is resistanceless when closed. All of the connecting wires may be considered to have zero resistance. All of the resistors shown are identical. The circuit contains an ideal voltmeter. The diagram shows all of the switches open. Below the diagram are eight different switch configurations for the circuit.

Rank these configurations in terms of the voltmeter reading. That is, put first the configuration for which the voltmeter gives the largest reading, and put last the configuration for which the voltmeter gives the smallest reading. If two or more configurations produce the same voltmeter reading, give these configurations equal ranking.

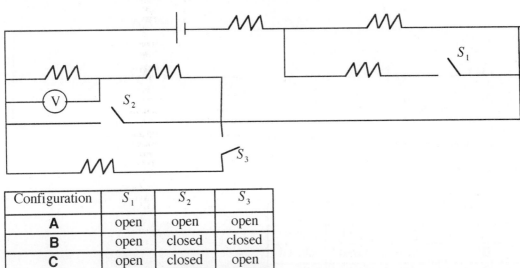

Configuration	S_1	S_2	S_3
A	open	open	open
B	open	closed	closed
C	open	closed	open
D	open	open	closed
E	closed	closed	closed
F	closed	open	open
G	closed	open	closed
H	closed	closed	open

Largest 1_____ 2_____ 3_____ 4_____ 5_____ 6_____ 7_____ 8_____ Smallest

Or, all configurations produce the same voltmeter reading. _____

Please carefully explain your reasoning.

How sure were you of your ranking? (circle one)

Basically Guessed   Sure   Very Sure

1   2   3   4   5   6   7   8   9   10

Shown below is a DC circuit that contains a number of switches. Each switch is resistanceless when closed. All of the connecting wires may be considered to have zero resistance. All of the resistors shown are identical. The circuit contains an ideal ammeter. The diagram shows all of the switches open. Below the diagram are eight different switch configurations for the circuit.

Rank these configurations in terms of the ammeter reading. That is, put first the configuration for which the ammeter gives the largest reading, and put last the configuration for which the ammeter gives the smallest reading. If two or more configurations produce the same ammeter reading, give these configurations equal ranking.

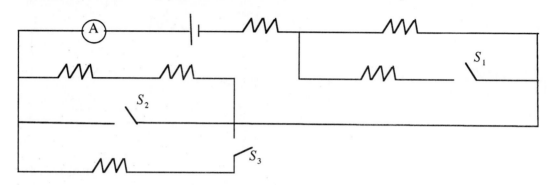

Configuration	S_1	S_2	S_3
A	open	open	open
B	open	closed	closed
C	open	closed	open
D	open	open	closed
E	closed	closed	closed
F	closed	open	open
G	closed	open	closed
H	closed	closed	open

Largest 1_____ 2_____ 3_____ 4_____ 5_____ 6_____ 7_____ 8_____ Smallest

Or, all configurations produce the same ammeter reading. _____

Please carefully explain your reasoning.

How sure were you of your ranking? (circle one)
Basically Guessed   Sure   Very Sure
1   2   3   4   5   6   7   8   9   10

In the circuit below, there are seven possible combinations of R_1, R_2, and R_3. Rank the combinations, in terms of the current measured by the ammeter, from highest to lowest.

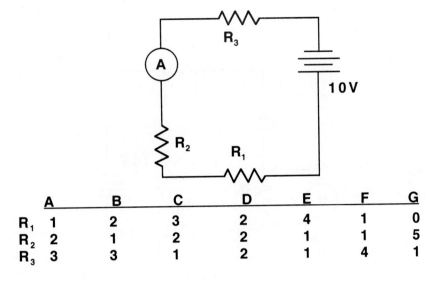

	A	B	C	D	E	F	G
R_1	1	2	3	2	4	1	0
R_2	2	1	2	2	1	1	5
R_3	3	3	1	2	1	4	1

Highest 1_____ 2_____ 3_____ 4_____ 5_____ 6_____ 7_____ Lowest

Or, all of the combinations have the same current. _____

Please carefully explain your reasoning.

How sure were you of your ranking? (circle one)
Basically Guessed   Very Sure
                                      Sure
1   2   3   4   5   6   7   8   9   10

In the circuit below, there are seven possible combinations of R_1, R_2, and R_3. Rank the combinations, on the basis of the reading on the voltmeter, from highest to lowest.

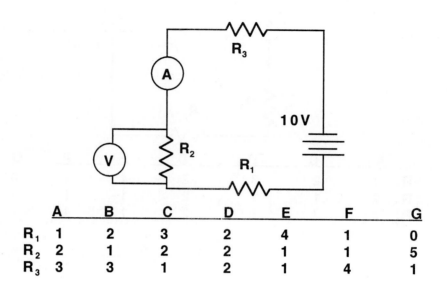

	A	B	C	D	E	F	G
R_1	1	2	3	2	4	1	0
R_2	2	1	2	2	1	1	5
R_3	3	3	1	2	1	4	1

Highest 1_____ 2_____ 3_____ 4_____ 5_____ 6_____ 7_____ Lowest

Or, all of the combinations have the same voltage. _____

Please carefully explain your reasoning.

How sure were you of your ranking? (circle one)

Basically Guessed   Sure   Very Sure

1   2   3   4   5   6   7   8   9   10

Resistive Circuit Concepts Diagnostic Test

The Resistive Circuit Concepts Diagnostic Test contains 20 ranking tasks designed to diagnose misconceptions about resistive circuits.

Instructions for students are provided as to how to answer and record answers to these ranking tasks questions.

Two types of student answer sheets are included:
1. Multiple choice answer sheets, which are useful for large classes because machine gradable answer forms can be used.
2. Write-in answer sheets, which require students to give a brief description of their reasoning. This one is better for classroom research and getting some insight into what students are thinking.

An instructor's key is provided, which lists correct answers as well as particular wrong answers that indicate the student's use of a wrong conception with codes for the wrong conceptions. Included in this section are descriptions of five categories of wrong conceptions. Each wrong conception has a Roman numeral code that is used on the instructor's answer key. If machine gradable answer forms are used, one can run them through several times, each time keyed to a particular wrong conception.

No numerical calculations are requested. All questions can be dealt with conceptually. The type of question used throughout is the ranking task. The following are examples of how you should mark your answer sheet. Place the greatest at the left and least toward the right. When two or more items are of the same "size," you must indicate so in the ranking. Notice how ties are dealt with in the examples.

1. Rank according to height.

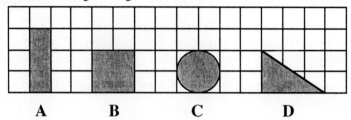

(on the answer sheet)

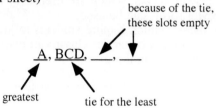

because of the tie,
these slots empty

A, BCD, ___, ___

greatest   tie for the least

1. Rank according to width.

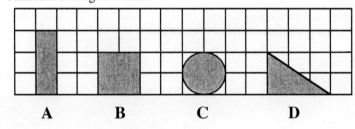

(on the answer sheet)   D , BC , A , ___

IMPORTANT:
In all of these batteries and bulbs questions:
        a) bulbs are identical
        b) treat the bulbs as ohmic resistors
        c) treat the wires as zero resistance connectors
        d) batteries are ideal (no internal resistance)

Resistive Circuit Concepts Diagnostic Test [180]

PART A (questions 1 – 5). Rank according to the brightness of bulb #1.

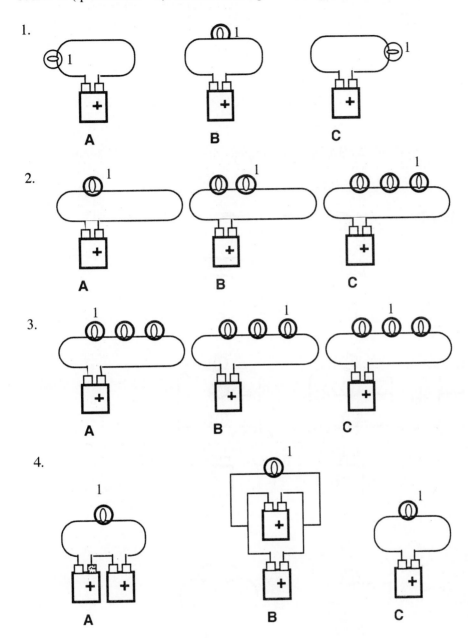

5.

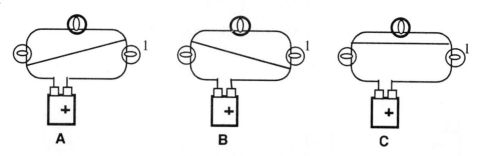

PART B (questions 6 – 9). Rank according to the ammeter reading (current). Assume that the ammeter has zero resistance.

6.

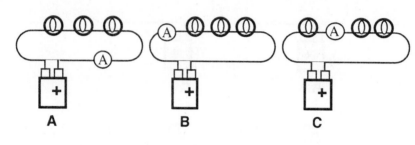

7.

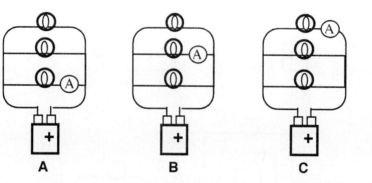

8.

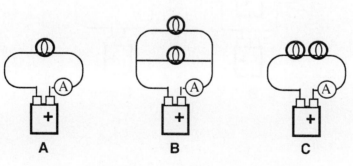

9.

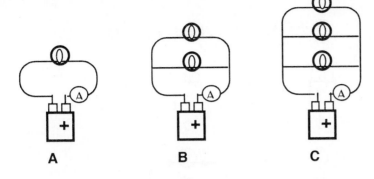

PART C (questions 10 – 16). Rank according to magnitude of difference in potential between points A and B.

10.

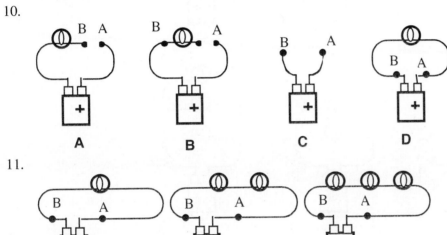

11.

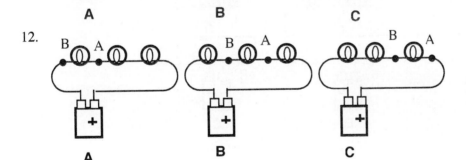

12.

13.

14.

15.

16.

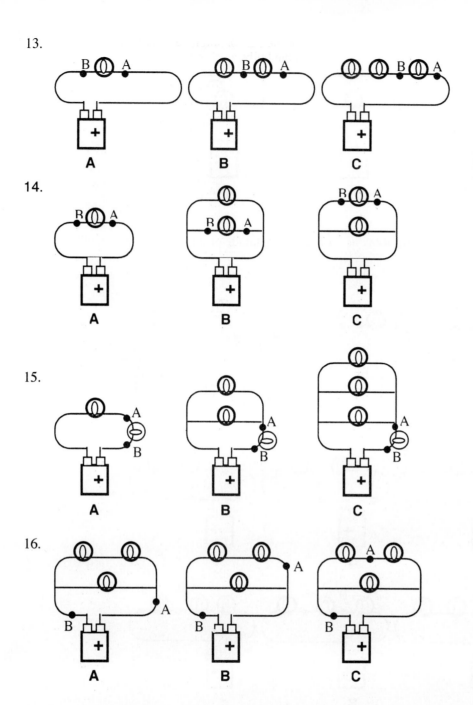

PART D (questions 17 – 20). Rank according to the size of single resistor that would be equivalent to the given group of resistors (treat the bulbs as ohmic resistors).

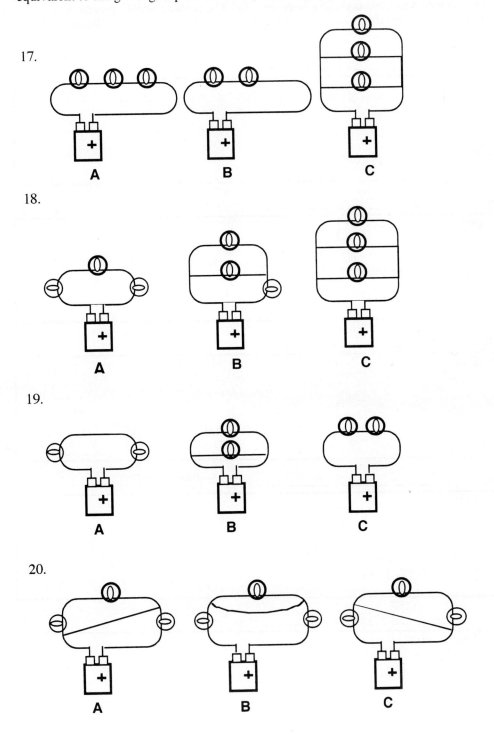

17.

A

B

C

18.

A

B

C

19.

A

B

C

20.

A

B

C

Resistive Circuit Concepts Diagnostic Test—Fill-In Answer Sheet[181]

Reminder: The answer ABC, means it is a three-way tie. The answer A, BC, means A is greatest while B and C tie for least. There is nothing in the final slot because of the tie.

1. Answer: _____ , _____ , _____

Briefly explain.

2. Answer: _____ , _____ , _____

Briefly explain.

3. Answer: _____ , _____ , _____

Briefly explain.

4. Answer: _____ , _____ , _____

Briefly explain.

5. Answer: _____ , _____ , _____

Briefly explain.

[181] D. Albers

6. Answer: _____ , _____ , _____

Briefly explain.

7. Answer: _____ , _____ , _____

Briefly explain.

8. Answer: _____ , _____ , _____

Briefly explain.

9. Answer: _____ , _____ , _____

Briefly explain.

10. Answer: _____ , _____ , _____

Briefly explain.

11. Answer: _____ , _____ , _____

Briefly explain.

12. Answer: _____ , _____ , _____

Briefly explain.

13. Answer: _____ , _____ , _____

Briefly explain.

14. Answer: _____ , _____ , _____

Briefly explain.

15. Answer: _____ , _____ , _____

Briefly explain.

16. Answer: _____ , _____ , _____

Briefly explain.

17. Answer: _____ , _____ , _____

Briefly explain.

18. Answer: _____ , _____ , _____

Briefly explain.

19. Answer: _____ , _____ , _____

Briefly explain.

20. Answer: _____ , _____ , _____

Briefly explain.

Resistive Circuit Concepts Diagnostic Test—Multiple Choice Answer Sheet [182]

Ask your instructor which you are to do: Mark a machine gradeable answer form, write on a separate sheet, or circle answers on this sheet. Reminder: The answer ABC, means it is a three-way tie. The answer A, BC, means A is greatest while B and C tie for least. The trailing comma indicates that the final slot is empty because of the tie.

1. a. B, C, A   b. C, B, A   c. ABC,   d. AC, B,

2. a. ABC,   b. A, B, C   c. C, B, A   d. none

3. a. ABC,   b. AC, B   c. B, C, A   d. none

4. a. B, AC,   b. A, BC,   c. AB, C,   d. none

5. a. A, C, B   b. ABC,   c. C, AB,   d. AB, C,

6. a. AB, C,   b. A, C, B   c. C, AB,   d. none

7. a. A, B, C   b. C, B, A   c. ABC,   d. none

8. a. AB, C,   b. B, A, C   c. ABC,   d. A, BC,

9. a. ABC,   b. A, BC,   c. A, B, C   d. none

10. a. BD, AC,   b. D, ABC,   c. ABCD,   d. ABD,C,

11. a. A, BC,   b. C, AB,   c. C, B, A   d. ABC,

12. a. ABC,   b. C, B, A   c. AC, B,   d. B, AC,

13. a. ABC,   b. A, B, C   c. A, BC,   d. none

14. a. ABC,   b. AB, C,   c. BC, A,   d. A, BC,

15. a. ABC,   b. A, B, C   c. A, BC,   d. none

16. a. A, B, C   b. C, B, A   c. AB, C   d. C, AB

17. a. AC, B,   b. C, A, B   c. A, B, C   d. none

18. a. A, B, C   b. B, AC,   c. ABC,   d. BC, A,

19. a. A, B, C   b. ABC,   c. B, AC,   d. AC, B,

20. a. ABC,   b. B, AC,   c. AC, B,   d. C, B, A

Resistive Circuit Concepts Diagnostic Test—Answer Key and Wrong Conception Code Key [183]

Question	Fill-In Correct	Fill-In Wrong	Multiple Choice Correct	Multiple Choice Wrong	Possible Wrong Conception
1	ABC,	C,B,A	c	b	Ia
2	A,B,C	ABC,	b	a	Ic
3	ABC,	B,C,A	a	c	Ib
4	A,BC,	B,AC,	b	a	IIIa
5	A,C,B	ABC,	a	b	IV
6	ABC,	A,C,B	d	b	Ia or Ib
7	ABC,	A,B,C	c	a	Ia or Ib
8	B,A,C	ABC,	b	c	Ic or IIIb
8	B,A,C	A,BC,	b	d	V
9	C,B,A	ABC,	d	a	Ic or IIIb
9	C,B,A	A,B,C	d	c	V
10	ABCD,	D,ABC,	c	b	IIa or IIb
11	ABC,	C,B,A	d	c	III plus V = IR (case C has largest R)
12	ABC,	C,B,A	a	b	I plus V = IR (case A has downstream I
13	A,B,C	ABC,	b	a	Ic plus V = IR
14	ABC,	A,BC,	a	d	IIIb plus V = IR
15	C,B,A	A,B,C	d	b	V plus V = IR
16	AB,C	any	c	a,b,d	(Student doesn't know that you can get from A to B along a path consisting only of zero resistance connectors, then V at A equals V at B so $\Delta V = 0$)
17	A,B,C	AC,B,	c	a	V
18	A,B,C	ABC,	a	c	V
19	AC,B,	ABC,	d	b	V
20	B,AC,	ABC,	b	a	IV plus v

Ia   Whatever lights the bulbs grows weaker downstream.

Ib   Whatever lights the bulbs is partially used up or weakened by upstream bulbs.

Ic   The leading bulb always has the same brightness regardless of what downstream bulbs exist.

IIa   Current, voltage, and energy are all sort of the same stuff. They flow around the circuit and into bulbs to light them. If the circuit is broken, there is no flow . . . no current, no voltage, and no energy.

IIb   When a circuit is broken, I = 0. Therefore, V = I R = 0 R = 0. That is, there is zero potential difference between points opposite the break. (This is an illegal use of Ohm's law.) Example:

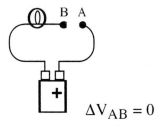

$$\Delta V_{AB} = 0$$

IIIa Batteries are fixed-current devices. Each battery in series puts out the current it would alone, so the bulb gets only that fixed current. When the batteries are in parallel, the fixed current from each battery joins to form a double current to light the bulb.

IIIb Batteries are fixed-current devices. When there is only one path from the battery the through bulb, the bulb gets that fixed current. When there are two bulbs and two paths (bulbs in parallel), each bulb gets half of that fixed current.

IV   A jumper wire has essentially R = 0. Therefore, V = I R = I O = 0. No voltage suggests no effect, so disregard the jumper wire's presence (another illegal use of Ohm's law).

V   The more bulbs, the greater the obstacle, hence, the smaller the battery current – regardless of how the bulbs are configured.

Magnetism and Electromagnetism Ranking Tasks

Moving charged particles are released with a velocity (details listed below) at the point P (2 m, 2 m) in a large region of space which has a uniform magnetic field in the +*x* direction. All these particles have the same mass, and they are released individually into this field.

Rank from greatest to least the magnitude of the initial acceleration of these charged particles as they are released from P.

Case	Charge	Speed	Direction
A	5 mC	3 m/s	+*x*
B	5 mC	3 m/s	-*x*
C	5 mC	3 m/s	+*y*
D	5 mC	3 m/s	-*y*
E	-10 mC	3 m/s	+*y*
F	+10 mC	3 m/s	-*y*
G	-10 mC	5 m/s	+*y*
H	+10 mC	5 m/s	-*y*

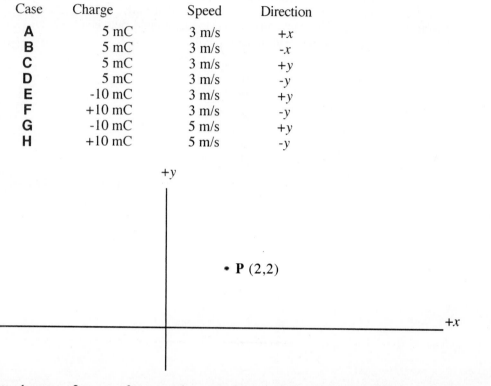

Greatest   1 _____ 2 _____ 3 _____ 4 _____ 5 _____ 6 _____ 7 _____ 8 _____ Least

Or, the magnitude of the initial acceleration will be the same for all of these (but not zero). ___

Or, the magnitude of the initial acceleration will be zero for all of these. ____

Please carefully explain your reasoning.

How sure were you of your ranking? (circle one)

Basically Guessed   Sure   Very Sure

1   2   3   4   5   6   7   8   9   10

Moving Charges in Uniform Magnetic Field—Change in Kinetic Energy[186]

Moving charged particles are released with a velocity (details listed below) at the point P (2 m, 2 m) in a large region of space which has a uniform magnetic field in the +x direction. All these particles have the same mass, and they are released individually into this field.

Rank from greatest to least the change in kinetic energy of these charged particles after they have traveled 1 cm from P.

Case	Charge	Speed	Direction
A	5 mC	3 m/s	+x
B	5 mC	3 m/s	-x
C	5 mC	3 m/s	+y
D	5 mC	3 m/s	-y
E	-10 mC	3 m/s	+y
F	+10 mC	3 m/s	-y
G	-10 mC	5 m/s	+y
H	+10 mC	5 m/s	-y

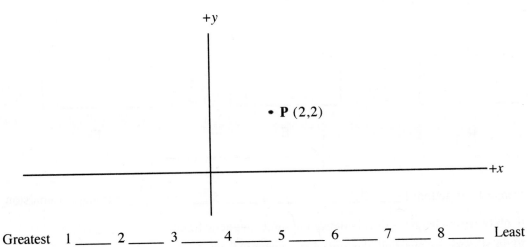

Greatest 1 _____ 2 _____ 3 _____ 4 _____ 5 _____ 6 _____ 7 _____ 8 _____ Least

Or, the change in kinetic energy will be the same for all of these (but not zero). _____

Or, the change in kinetic energy will be zero for all of these. _____

Please carefully explain your reasoning.

How sure were you of your ranking? (circle one)

Basically Guessed				Sure				Very Sure	
1	2	3	4	5	6	7	8	9	10

Charges Near Magnets—Magnetic Force[187]

The figures below show electrically charged particles sitting at rest near the poles of permanent magnets. All of the permanent magnets are the same strength. The magnitudes and signs of the electric charges are shown in the circles, which represent the particles. The zero values indicate that the particle is actually electrically neutral.

Rank these situations, from strongest attraction to strongest repulsion, on the basis of the magnetic force exerted by the magnet on the charge. As stated, if the interaction is actually a repulsion rather than an attraction, it would be ranked lower than any attraction.

Strongest Attraction 1_____ 2_____ 3_____ 4_____ 5_____ 6_____ Strongest Repulsion

Or, all of these charges will experience the same magnetic force. _____

Or, all of these charges will experience no magnetic force. _____

Please carefully explain your reasoning.

How sure were you of your ranking? (circle one)

Basically Guessed				Sure				Very Sure	
1	2	3	4	5	6	7	8	9	10

Pairs of Long Current Carrying Wires—Magnetic Field[188]

Shown below are several situations with two wires with currents of 2 A flowing out of (•) or into (X) the page. Rank the magnetic field at point **P** from greatest to least. Count a downward pointing field as negative and upward as positive. You may assume that each wire or point of interest is separated by the same distance from the adjacent items and that each situation is independent of all others.

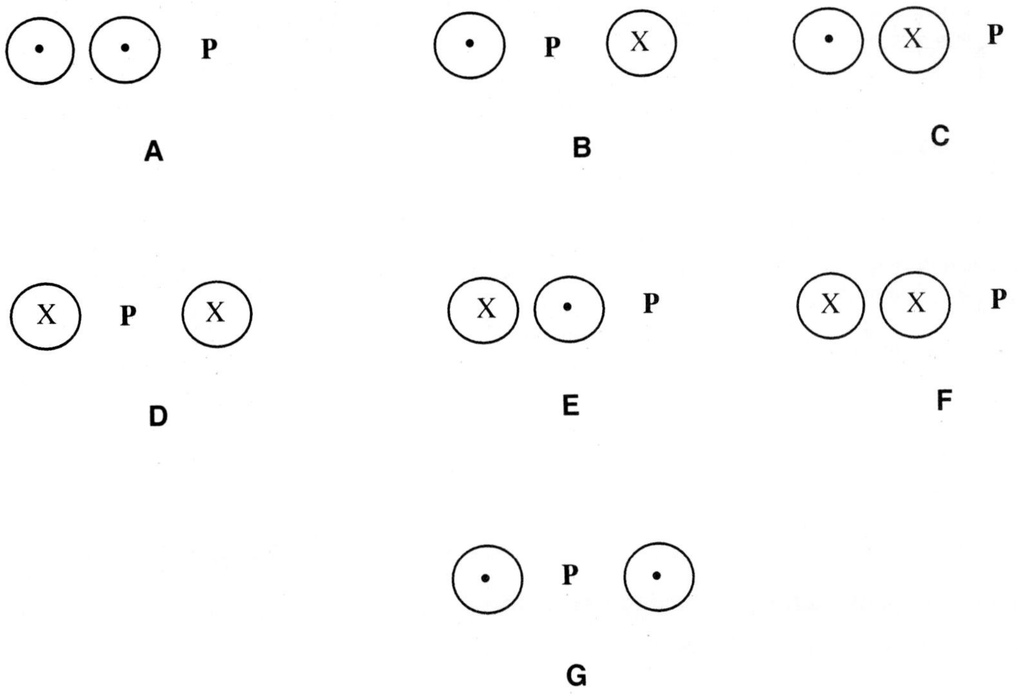

A

B

C

D

E

F

G

Greatest 1 _____ 2 _____ 3 _____ 4 _____ 5 _____ 6 _____ 7 _____ Least

Or, the magnetic field is the same in all cases. _____

Or, the magnetic field is zero in all cases. _____

Please carefully explain your reasoning.

How sure were you of your ranking? (Circle one)

Guessed   Sure   Very Sure
1   2   3   4   5   6   7   8   9   10

---

[188] C. Davies, B. Diesslin, M. Nelson, G. Shepherd
Ranking Task Exercises in Physics

Electricity and Magnetism

Pairs of Equal Current Electromagnets—Force[189]

Eight pairs of electromagnets are shown below. The same current flows in each case. The pairs of electromagnets are separated by the same distance, and they have the same length and diameter. Carefully observe the orientation of the coil and direction of current flow.

Rank the forces between the magnets from the most attractive to the most repulsive.

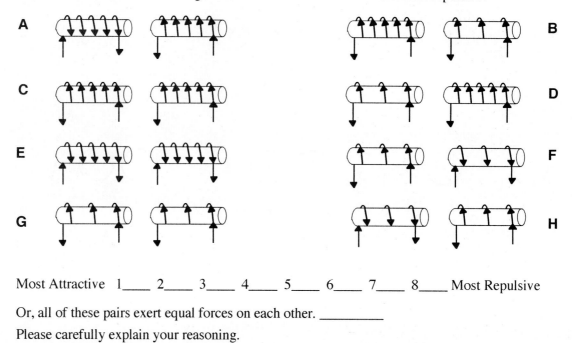

Most Attractive   1____ 2____ 3____ 4____ 5____ 6____ 7____ 8____ Most Repulsive

Or, all of these pairs exert equal forces on each other. _____

Please carefully explain your reasoning.

How sure were you of your ranking? (circle one)
Basically Guessed   Sure   Very Sure
1   2   3   4   5   6   7   8   9   10

[189] B. Emerson, C. Hieggelke

Pairs of Equal Current Electromagnets—Magnetic Field Between[190]

Eight pairs of electromagnets are shown below. The same current flows in each electromagnet. The pairs are separated by the same distance, and they have the same length and diameter. Carefully observe the orientation of the coil and direction of current flow in the electromagnets.

Rank the magnitudes of the magnetic fields at the midpoint between the electromagnets from the largest to the smallest.

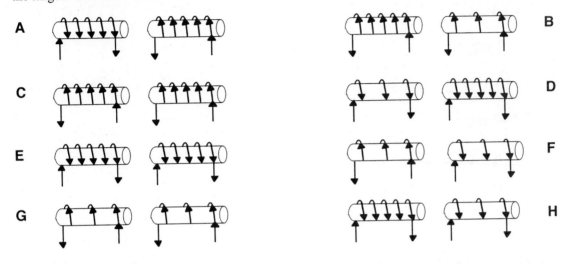

Largest   1____   2____   3____   4____   5____   6____   7____   8____   Smallest

Or, the magnitudes of the magnetic field are all the same.____

Please carefully explain your reasoning.

How sure were you of your ranking? (circle one)
Basically Guessed   Sure   Very Sure
1   2   3   4   5   6   7   8   9   10

[190] B. Emerson, C. Hieggelke
Ranking Task Exercises in Physics

Electricity and Magnetism

Unequal Current Electromagnets—Magnetic Field at Ends[191]

Eight electromagnets are shown below. The current in each electromagnet is indicated in the diagram as well as the number of turns. They have the same length and diameter. Carefully observe the orientation of the turns and direction of current flow in the coil.

Rank the magnetic field just outside the right end of the electromagnet from the largest to the smallest. (The direction to the right is positive, and –5 < -3.)

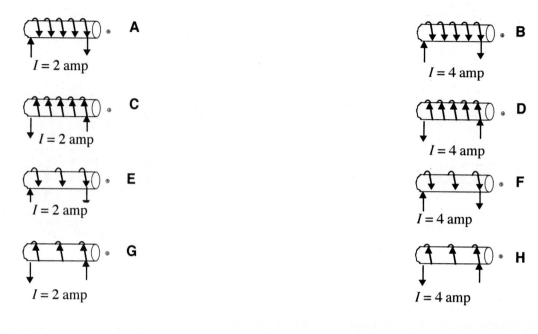

Largest 1_____ 2_____ 3_____ 4_____ 5_____ 6_____ 7_____ 8_____ Smallest

Or, the magnitude of the magnetic field is the same but not zero._____

Or, the magnitude of the magnetic field is zero._____

Please carefully explain your reasoning.

How sure were you of your ranking? (circle one)

Basically Guessed   Sure   Very Sure

1   2   3   4   5   6   7   8   9   10

[191] B. Emerson, C. Hieggelke

Eight pairs of electromagnets are shown below. The current in the left electromagnet is one amp and the current in the right one is two amps in each case. They are also separated by the same distance, and they have the same length and diameter. Carefully observe the orientation of the coil and direction of current flow.

Rank the magnetic fields at the midpoints between the electromagnets from the largest to the smallest. (The direction to the right is positive, and $-5 < -3$.)

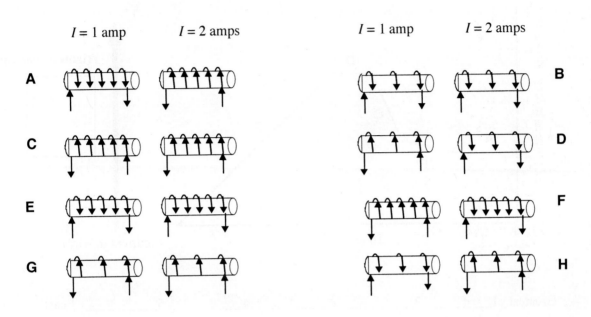

Largest 1_____ 2_____ 3_____ 4_____ 5_____ 6_____ 7_____ 8_____   Smallest

Or, the magnitude of the magnetic field is the same but not zero for all eight pairs. _____

Or, the magnitude of the magnetic field is zero for all eight pairs. _____

Please carefully explain your reasoning.

How sure were you of your ranking? (circle one)

Basically Guessed				Sure				Very Sure	
1	2	3	4	5	6	7	8	9	10

We have a long wire with a varying current, I, as indicated in the graph below. Near this wire is a square loop, also shown in the diagram below.

Rank the magnitude of the induced current, i, in the square loop from greatest to least for the different time intervals.

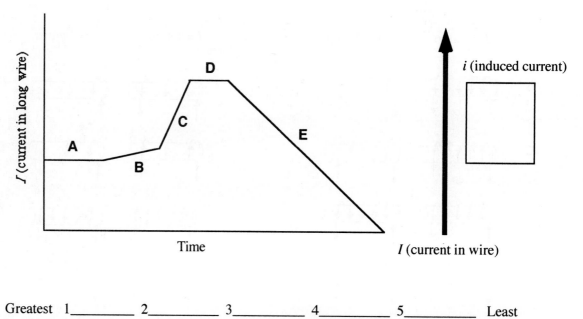

Greatest 1_____ 2_____ 3_____ 4_____ 5_____ Least

Or, all of the intervals produce the same induced current. _____

Please carefully explain your reasoning.

How sure were you of your ranking? (circle one)
Basically Guessed   Sure   Very Sure
1   2   3   4   5   6   7   8   9   10

Answer Key

Kinematics Ranking Tasks							1
Ball Motion Diagrams—Velocity I	ADF	BE	C				2
Ball Motion Diagrams—Acceleration I	ADF	BE	C				3
Ball Motion Diagrams—Velocity II	F	BE	C	AD			4
Ball Motion Diagrams—Acceleration II	F	C	BE	AD			5
Objects in Different Situations—Accelerations	E	F	A	BC	D		6
Vertical Model Rockets—Maximum Height	BF	AEG	CDH				7
Vertical Arrows—Maximum Height	C	AF	BD	EG	H		8
Position Time Graphs—Displacement	C	D	ABF	E			9
Cars—Change of Velocity	AE	D	BFG	C	H		10
Position Time Graphs—Average Speed	E	F	C	D	AB		11
Motion Diagrams—Displacement	DF	E	A	B	C		12
Motion Diagrams—Average Velocity	DF	E	A	B	C		13
People on Trains—Speed Relative to Ground	D	F	BE	AC			14
Force Ranking Tasks							**15**
Carts Moving along Horizontal Surface—String Tension	A	BE	C	DF			16
Carts Moving Along Horizontal Surface—Acceleration	DF	C	BE	A			17
Carts Moving along Horizontal Surface—Slowing Down	DF	C	BE	A			18
Two-Dimensional Forces on a Treasure Chest—Final Speed	AE	F	BD	C			19
Two-Dimensional Forces on a Treasure Chest—Acceleration	AE	F	BD	C			20
Arrows—Acceleration	All same (g)						21
Rocks Thrown Upward—Net Force	CE	BH	DG	AF			22
Model Rockets Moving Upward—Net Force	AD	CFG	BH	E			23
Blocks Attached to Fixed Objects—Rope Tension	E	ACDFG	B	H			24
Ball Motion Diagram—Net Force	ADF	BE	C				25
Force Acceleration Graphs—Mass	F	C	AD	B	E		26
Two Different Blocks and a Pulley—Tension	E	F	C	D	B	A	27
Ropes Pulling Boxes—Acceleration	A	BC	DEF				28
Ropes Pulling Boxes—Rope Tension	D	BE	ACF				29
Two Different Blocks and a Pulley—Net Force	F	D	E	BC	A		30
Moving Car and Boat Trailer—Force Difference	All same						31
Accelerating Car and Boat Trailer—Force Difference	All same						32
Car and Boat Trailer on an Incline—Force Difference	All same						33
Forces on Objects on Smooth Surfaces—Velocity Change	B	A	CF	D	E		34
Forces on Objects on Smooth Surfaces—Speed Change	BE	AD	CF				35
Forces on Objects on Rough Surfaces—Velocity Change	B	A	CF	D	E		36
Forces on Objects on Rough Surfaces—Speed Change	All same						37
Person in an Elevator Moving Upward—Scale Weight	AE	CDF	B	GH			38
Person in an Elevator Moving Downward—Scale Weight	AE	CDF	B	GH			39
Two Blocks at Rest—Force Difference	All same - 0						40
Two Moving Blocks—Force Difference	All same - 0						41
Two Accelerating Blocks—Force Difference	All same - 0						42
Horizontal Arrows at Different Distances—Force	All same - 0						43
Horizontal Arrows at Different Times—Force	All same - 0						44
Horizontal Arrows at Different Distances and Times—Force	All same - 0						45
Projectile and Other Two Dimensional Motion Ranking Tasks							**46**
Water Over a Waterfall—Time to Reach Ground	ABC	DEF					47
Horizontal Arrows—Time to Hit Ground	BC	DGH	AEF				48
Rifle Shots—Time to Hit Ground	BC	DGH	AEF				49
Toy Trucks Rolling Off Tables—Time in Air	BC	DGH	AEF				50
Spheres Thrown Horizontally Off Cliffs—Time to Hit Ground	BC	DGH	AEF				51
Arrows—Maximum Heights	CF	BD	AEG	H			52
Rock Throw—Maximum Heights	CDF	BG	AH	E			53
Model Rockets Fired at an Angle—Horizontal Speed at Top	BEG	AFH	CD				54
Cannon Shots—Acceleration at the Top	All same						55
Carts on Incline—Height After Leaving Incline	AD	CE	H	BFG			56
Projectile—Horizontal Distance	EF	CD	AB				57
Projectile—Time in Air	All same						58
Work-Energy Ranking Tasks							**59**
Boxcars and Ropes—Stopping Force in Same Distance	C	F	D	E	B	A	60
Cars and Barriers—Stopping Force in Same Distance I	C	H	E	D	G	B	F	A	61
Cars and Barriers—Stopping Distance with the Same Force	G	H	C	F	B	D	E	A	62
Cars—Work Done in Change of Velocity	E	AG	CD	B	FH		63
Bouncing Cart—Change in Kinetic Energy	EF	BCD	A				64
Bouncing Cart—Work Done by the Barrier	EF	BCD	A				65
Bouncing Cart—Work Done on the Barrier	A	BCD	EF				66

Ranking Task Exercises in Physics   215

Answer Key

Ranking Task Exercises in Physics
Prentice Hall, Inc.
Macintosh System Requirements:
Apple Power Macintosh computer
Apple System Software version 7.1.2 or later
4.5 MB of available RAM (6.5 MB recommended)
8 MB of available hard-disk space
Windows System Requirements:
i486 or Pentium processor
Windows 95/98/NT 4.0X
8 MB RAM on Windows 95/98 system (16 MB recommended)
16 MB RAM on Windows NT system (24 MB recommended)

LICENSE AGREEMENT

YOU SHOULD CAREFULLY READ THE FOLLOWING TERMS AND CONDITIONS BEFORE BREAKING THE SEAL ON THE PACKAGE. AMONG OTHER THINGS, THIS AGREEMENT LICENSES THE ENCLOSED SOFTWARE TO YOU AND CONTAINS WARRANTY AND LIABILITY DISCLAIMERS. BY BREAKING THE SEAL ON THE PACKAGE, YOU ARE ACCEPTING AND AGREEING TO THE TERMS AND CONDITIONS OF THIS AGREEMENT. IF YOU DO NOT AGREE TO THE TERMS OF THIS AGREEMENT, DO NOT BREAK THE SEAL. YOU SHOULD PROMPTLY RETURN THE PACKAGE UNOPENED.

LICENSE.
Subject to the provisions contained herein, Prentice-Hall, Inc. ("PH") hereby grants to you a non-exclusive, non-transferable license to use the object code version of the computer software product ("Software") contained in the package on a single computer of the type identified on the package.

SOFTWARE AND DOCUMENTATION.
PH shall furnish the Software to you on media in machine-readable object code form and may also provide the standard documentation ("Documentation") containing instructions for operation and use of the Software.

LICENSE TERM AND CHARGES.
The term of this license commences upon delivery of the Software to you and is perpetual unless earlier terminated upon default or as otherwise set forth herein.

TITLE.
Title, and ownership right, and intellectual property rights in and to the Software and Documentation shall remain in PH and/or in suppliers to PH of programs contained in the Software. The Software is provided for your own internal use under this license. This license does not include the right to sublicense and is personal to you and therefore may not be assigned (by operation of law or otherwise) or transferred without the prior written consent of PH. You acknowledge that the Software in source code form remains a confidential trade secret of PH and/or its suppliers and therefore you agree not to attempt to decipher or decompile, modify, disassemble, reverse engineer or prepare derivative works of the Software or develop source code for the Software or knowingly allow others to do so. Further, you may not copy the Documentation or other written materials accompanying the Software.

UPDATES.
This license does not grant you any right, license, or interest in and to any improvements, modifications, enhancements, or updates to the Software and Documentation. Updates, if available, may be obtained by you at PH's then current standard pricing, terms, and conditions.

LIMITED WARRANTY AND DISCLAIMER.
PH warrants that the media containing the Software, if provided by PH, is free from defects in material and workmanship under normal use for a period of sixty (60) days from the date you purchased a license to it.

THIS IS A LIMITED WARRANTY AND IT IS THE ONLY WARRANTY MADE BY PH. THE SOFTWARE IS PROVIDED 'AS IS' AND PH SPECIFICALLY DISCLAIMS ALL WARRANTIES OF ANY KIND, EITHER EXPRESS OR IMPLIED, INCLUDING, BUT NOT LIMITED TO, THE IMPLIED WARRANTY OF MERCHANTABILITY AND FITNESS FOR A PARTICULAR PURPOSE. FURTHER, COMPANY DOES NOT WARRANT, GUARANTY OR MAKE ANY REPRESENTATIONS REGARDING THE USE, OR THE RESULTS OF THE USE, OF THE SOFTWARE IN TERMS OF CORRECTNESS, ACCURACY, RELIABILITY, CURRENTNESS, OR OTHERWISE AND DOES NOT WARRANT THAT THE OPERATION OF ANY SOFTWARE WILL BE UNINTERRUPTED OR ERROR FREE. COMPANY EXPRESSLY DISCLAIMS ANY WARRANTIES NOT STATED HEREIN. NO ORAL OR WRITTEN INFORMATION OR ADVICE GIVEN BY PH, OR ANY PH DEALER, AGENT, EMPLOYEE OR OTHERS SHALL CREATE, MODIFY OR EXTEND A WARRANTY OR IN ANY WAY INCREASE THE SCOPE OF THE FOREGOING WARRANTY, AND NEITHER SUBLICENSEE OR PURCHASER MAY RELY ON ANY SUCH INFORMATION OR ADVICE. If the media is subjected to accident, abuse, or improper use; or if you violate the terms of this Agreement, then this warranty shall immediately be terminated. This warranty shall not apply if the Software is used on or in conjunction with hardware or programs other than the unmodified version of hardware and programs with which the Software was designed to be used as described in the Documentation.

LIMITATION OF LIABILITY.
Your sole and exclusive remedies for any damage or loss in any way connected with the Software are set forth below. UNDER NO CIRCUMSTANCES AND UNDER NO LEGAL THEORY, TORT, CONTRACT, OR OTHERWISE, SHALL PH BE LIABLE TO YOU OR ANY OTHER PERSON FOR ANY INDIRECT, SPECIAL, INCIDENTAL, OR CONSEQUENTIAL DAMAGES OF ANY CHARACTER INCLUDING, WITHOUT LIMITATION, DAMAGES FOR LOSS OF GOODWILL, LOSS OF PROFIT, WORK STOPPAGE, COMPUTER FAILURE OR MALFUNCTION, OR ANY AND ALL OTHER COMMERCIAL DAMAGES OR LOSSES, OR FOR ANY OTHER DAMAGES EVEN IF PH SHALL HAVE BEEN INFORMED OF THE POSSIBILITY OF SUCH DAMAGES, OR FOR ANY CLAIM BY ANY OTHER PARTY. PH'S THIRD PARTY PROGRAM SUPPLIERS MAKE NO WARRANTY, AND HAVE NO LIABILITY WHATSOEVER, TO YOU. PH's sole and exclusive obligation and liability and your exclusive remedy shall be: upon PH's election, (i) the replacement of your defective media; or (ii) the repair or correction of your defective media if PH is able, so that it will conform to the above warranty; or (iii) if PH is unable to replace or repair, you may terminate this license by returning the Software. Only if you inform PH of your problem during the applicable warranty period will PH be obligated to honor this warranty. You may contact PH to inform PH of the problem as follows:

SOME STATES OR JURISDICTIONS DO NOT ALLOW THE EXCLUSION OF IMPLIED WARRANTIES OR LIMITATION OR EXCLUSION OF CONSEQUENTIAL DAMAGES, SO THE ABOVE LIMITATIONS OR EXCLUSIONS MAY NOT APPLY TO YOU. THIS WARRANTY GIVES YOU SPECIFIC LEGAL RIGHTS AND YOU MAY ALSO HAVE OTHER RIGHTS WHICH VARY BY STATE OR JURISDICTION.

MISCELLANEOUS.
If any provision of this Agreement is held to be ineffective, unenforceable, or illegal under certain circumstances for any reason, such decision shall not affect the validity or enforceability (i) of such provision under other circumstances or (ii) of the remaining provisions hereof under all circumstances and such provision shall be reformed to and only to the extent necessary to make it effective, enforceable, and legal under such circumstances. All headings are solely for convenience and shall not be considered in interpreting this Agreement. This Agreement shall be governed by and construed under New York law as such law applies to agreements between New York residents entered into and to be performed entirely within New York, except as required by U.S. Government rules and regulations to be governed by Federal law.

YOU ACKNOWLEDGE THAT YOU HAVE READ THIS AGREEMENT, UNDERSTAND IT, AND AGREE TO BE BOUND BY ITS TERMS AND CONDITIONS. YOU FURTHER AGREE THAT IT IS THE COMPLETE AND EXCLUSIVE STATEMENT OF THE AGREEMENT BETWEEN US THAT SUPERSEDES ANY PROPOSAL OR PRIOR AGREEMENT, ORAL OR WRITTEN, AND ANY OTHER COMMUNICATIONS BETWEEN US RELATING TO THE SUBJECT MATTER OF THIS AGREEMENT.

U.S. GOVERNMENT RESTRICTED RIGHTS.
Use, duplication or disclosure by the Government is subject to restrictions set forth in subparagraphs (a) through (d) of the Commercial Computer-Restricted Rights clause at FAR 52.227-19 when applicable, or in subparagraph (c) (1) (ii) of the Rights in Technical Data and Computer Software clause at DFARS 252.227-7013, and in similar clauses in the NASA FAR Supplement.